LONDON MATHEMATICAL SOCIETY STUDEN'

Managing editor: Professor C.M. Series, Mathematica
University of Warwick, Coventry CV4 7AL, United K

3   Local fields, J.W.S. CASSELS
4   An introduction to twistor theory: Second edition, S.A. HUGGETT & K.P. TOD
5   Introduction to general relativity, L.P. HUGHSTON & K.P. TOD
7   The theory of evolution and dynamical systems, J. HOFBAUER & K. SIGMUND
8   Summing and nuclear norms in Banach space theory, G.J.O. JAMESON
9   Automorphisms of surfaces after Nielsen and Thurston, A. CASSON & S. BLEILER
11 Spacetime and singularities, G. NABER
12 Undergraduate algebraic geometry, MILES REID
13 An introduction to Hankel operators, J.R. PARTINGTON
15 Presentations of groups: Second edition, D.L. JOHNSON
17 Aspects of quantum field theory in curved spacetime, S.A. FULLING
18 Braids and coverings: selected topics, VAGN LUNDSGAARD HANSEN
19 Steps in commutative algebra, R.Y. SHARP
20 Communication theory, C.M. GOLDIE & R.G.E. PINCH
21 Representations of finite groups of Lie type, FRANÇOIS DIGNE & JEAN MICHEL
22 Designs, graphs, codes, and their links, P.J. CAMERON & J.H. VAN LINT
23 Complex algebraic curves, FRANCES KIRWAN
24 Lectures on elliptic curves, J.W.S. CASSELS
26 An Introduction to the theory of L-functions and Eisenstein series, H. HIDA
27 Hilbert Space: compact operators and the trace theorem, J.R. RETHERFORD
28 Potential theory in the complex plane, T. RANSFORD
29 Undergraduate commutative algebra, M. REID
31 The Laplacian on a Riemannian manifold, S. ROSENBERG
32 Lectures on Lie Groups and Lie Algebras, R. CARTER, G. SEGAL
    & I. MACDONALD
33 A primer of algebraic D-modules, S.C. COUTINHO
34 Complex algebraic surfaces, A. BEAUVILLE
35 Young tableaux, W. FULTON
37 A mathematical introduction to wavelets, P. WOJTASCZYK
38 Harmonic maps, loop groups and integrable systems, M. GUEST
39 Set theory for the working mathematician, K. CIESIELSKI
40 Ergodic theory and dynamical systems, M. POLLICOTT & M. YURI
41 The algorithmic resolution of diophantine equations, N.P. SMART
42 Equilibrium states in ergodic theory, G. KELLER
44 Classical invariant theory, P. OLVER
45 Permutation groups, P.J. CAMERON

London Mathematical Society Student Texts 32

Lectures on Lie Groups and Lie Algebras

Roger Carter
University of Warwick

Graeme Segal
University of Cambridge

Ian Macdonald
Queen Mary and Westfield College, London

PUBLISHED BY THE PRESS SYNDICATE OF THE UNIVERSITY OF CAMBRIDGE
The Pitt Building, Trumpington Street, Cambridge, United Kingdom

CAMBRIDGE UNIVERSITY PRESS
The Edinburgh Building, Cambridge CB2 2RU, UK
40 West 20th Street, New York, NY 10011-4211, USA
477 Williamstown Road, Port Melbourne, VIC 3207, Australia
Ruiz de Alarcón 13, 28014 Madrid, Spain
Dock House, The Waterfront, Cape Town 8001, South Africa

http://www.cambridge.org

© Cambridge University Press 1995

This book is in copyright. Subject to statutory exception
and to the provisions of relevant collective licensing agreements,
no reproduction of any part may take place without
the written permission of Cambridge University Press.

First published 1995
Reprinted 1999, 2002

Printed in the United Kingdom at the University Press, Cambridge

Typeface Monotype Times 11/13.5pt.   *System* LATEX [EPC]

A catalogue record for this book is available from the British Library

ISBN 0 521 49579 2 hardback
ISBN 0 521 49922 4 paperback

Contents

Foreword M. J. Taylor		*page* vii
Lie Algebras and Root Systems R. W. Carter		1
Preface		3
1	**Introduction to Lie algebras**	5
1.1	Basic concepts	5
1.2	Representations and modules	7
1.3	Special kinds of Lie algebra	8
1.4	The Lie algebras $sl_n(\mathbb{C})$	10
2	**Simple Lie algebras over $\mathbb{C}$**	12
2.1	Cartan subalgebras	12
2.2	The Cartan decomposition	13
2.3	The Killing form	15
2.4	The Weyl group	16
2.5	The Dynkin diagram	18
3	**Representations of simple Lie algebras**	25
3.1	The universal enveloping algebra	25
3.2	Verma modules	26
3.3	Finite dimensional irreducible modules	27
3.4	Weyl's character and dimension formulae	29
3.5	Fundamental representations	32
4	**Simple groups of Lie type**	36
4.1	A Chevalley basis of g	36
4.2	Chevalley groups over an arbitrary field	38
4.3	Finite Chevalley groups	39
4.4	Twisted groups	41
4.5	Suzuki and Ree groups	43
4.6	Classification of finite simple groups	44

Lie Groups *Graeme Segal* 45

Introduction 47
1. Examples 49
2. SU_2, SO_3, and $SL_2\mathbb{R}$ 53
3. Homogeneous spaces 59
4. Some theorems about matrices 63
5. Lie theory 69
6. Representation theory 82
7. Compact groups and integration 85
8. Maximal compact subgroups 89
9. The Peter-Weyl theorem 91
10. Functions on $\mathbb{R}^n$ and S^{n-1} 100
11. Induced representations 104
12. The complexification of a compact group 108
13. The unitary and symmetric groups 110
14. The Borel-Weil theorem 115
15. Representations of non-compact groups 120
16. Representations of $SL_2\mathbb{R}$ 124
17. The Heisenberg group 128

Linear Algebraic Groups *I. G. Macdonald* 133

Preface 135
Introduction 137
1. Affine algebraic varieties 139
2. Linear algebraic groups: definition and elementary properties 146
Interlude 154
3. Projective algebraic varieties 157
4. Tangent spaces. Separability 162
5. The Lie algebra of a linear algebraic group 166
6. Homogeneous spaces and quotients 172
7. Borel subgroups and maximal tori 177
8. The root structure of a linear algebraic group 182
Notes and references 186
Bibliography 187
Index 189

Foreword

This book consists of notes based on the three introductory lecture courses given at the LMS-SERC Instructional Conference on Lie theory and algebraic groups held at Lancaster University in September 1993: Lie Algebras by Roger Carter; Lie Groups by Graeme Segal; algebraic groups by Ian Macdonald.

The aim of the course was to provide an introduction to this important area of mathematics for postgraduate students who had no previous specialised knowledge. Discussions with the students at the end of the meeting suggested that the conference had been extremely successful; it then seemed desirable to further impose on the lecturers by asking them to write-up their lectures, in order that future generations of students could also benefit from their efforts.

All three lecturers adopted the same approach of providing a crisp, fast-moving, clear introduction, while at the same time taking care to indicate more advanced material, so as to give the full flavour of the subject. It is clear, from both the lectures and the written account, that a substantial effort was made to ensure a coherent and well-harmonised presentation of these three highly interrelated themes.

The general intention of the new series of LMS-SERC Instructional Conferences is to provide postgraduate students with the opportunity to learn important mainstream core mathematics, which they might not otherwise meet. Lie theory and algebraic groups seemed to be a natural first choice, since they are a central mathematical crossroads, which relate to a host of important areas such as group theory, number theory, algebraic geometry, differential geometry, topology, particle physics and strings; indeed, a knowledge of algebraic groups and Lie theory can be quite crucial in making significant progress in many aspects of these related areas.

Finally, on behalf of the LMS, I should like to express my deepest gratitude to the three authors for not just accepting to give their lectures and then write them up, but also for carrying out their alloted tasks with such infectious enthusiasm; extra special thanks go to Ian Macdonald for sage advice in the initial planning of the meeting. It is also a pleasure to thank both Roger Astley and David Tranah for their help and cooperation in producing these notes, which will be a very valuable contribution to the mathematical community.

Martin Taylor
UMIST
Manchester

Lie Algebras and Root Systems

R.W. Carter

Contents
Lie Algebras and Root Systems

Preface		3
1	**Introduction to Lie algebras**	5
1.1	Basic concepts	5
1.2	Representations and modules	7
1.3	Special kinds of Lie algebra	8
1.4	The Lie algebras $sl_n(\mathbb{C})$	10
2	**Simple Lie algebras over $\mathbb{C}$**	12
2.1	Cartan subalgebras	12
2.2	The Cartan decomposition	13
2.3	The Killing form	15
2.4	The Weyl group	16
2.5	The Dynkin diagram	18
3	**Representations of simple Lie algebras**	25
3.1	The universal enveloping algebra	25
3.2	Verma modules	26
3.3	Finite dimensional irreducible modules	27
3.4	Weyl's character and dimension formulae	29
3.5	Fundamental representations	32
4	**Simple groups of Lie type**	36
4.1	A Chevalley basis of g	36
4.2	Chevalley groups over an arbitrary field	38
4.3	Finite Chevalley groups	39
4.4	Twisted groups	41
4.5	Suzuki and Ree groups	43
4.6	Classification of finite simple groups	44

Preface

The following notes on Lie Algebras and Root Systems follow fairly closely the lectures I gave on this subject at the Lancaster meeting, although more detail has been included in a number of places. The aim has been to give an outline of the main ideas involved in the structure and representation theory of the simple Lie algebras over $\mathbb{C}$, and the construction of the corresponding groups of Lie type over an arbitrary field.

It has not been possible to give all the proofs in detail, and so interested readers are encouraged to consult books in which more complete information is given. The following books are particularly recommended.

J. E. Humphreys, Introduction to Lie Algebras and Representation Theory, Graduate Texts in Mathematics 9 (1972) Springer.

N. Jacobson, Lie Algebras. Interscience Publishers, J. Wiley, New York (1962).

R. W. Carter, Simple Groups of Lie Type, Wiley Classics Library Edition (1989), J. Wiley, New York.

1
Introduction to Lie algebras

1.1 Basic concepts

A Lie algebra is a vector space g over a field F on which a multiplication

$$g \times g \to g$$
$$x, y \to [xy]$$

is defined satisfying the axioms:

(i) $[xy]$ is linear in x and in y.
(ii) $[xx] = 0$ for all $x \in g$.
(iii) $[[xy]z] + [[yz]x] + [[zx]y] = 0$ for all $x, y, z \in g$.

Property (iii) is called the Jacobi identity.

We note that the multiplication is not associative, i.e., it is not true in general that $[[xy]z] = [x[yz]]$. It is therefore essential to include the Lie brackets in products of elements.

For any pair of elements $x, y \in g$ we have

$$[x + y, x + y] = [xx] + [xy] + [yx] + [yy].$$

We also know that

$$[xx] = 0, \quad [yy] = 0, \quad [x + y, x + y] = 0.$$

It follows that $[yx] = -[xy]$ for all $x, y \in g$. Thus multiplication in a Lie algebra is anticommutative.

Lie algebras can be obtained from associative algebras by the following method. Let A be an associative algebra, i.e., a vector space with a bilinear associative multiplication xy. Then we may obtain a Lie algebra $[A]$ by redefining the multiplication on A. We define $[xy] = xy - yx$. It is clear that $[xy]$ is linear in x and in y and that $[xx] = 0$. We also have

$$\begin{aligned}{}[[xy]z] &= (xy - yx)z - z(xy - yx) \\ &= xyz - yxz - zxy + zyx.\end{aligned}$$

It follows that

$$\begin{aligned}&[[xy]z] + [[yz]x] + [[zx]y] \\ =\ & xyz - yxz - zxy + zyx \\ &+yzx - zyx - xyz + xzy \\ &+zxy - xzy - yzx + yxz \\ =\ & 0,\end{aligned}$$

so that the Jacobi identity is satisfied.

Let g_1, g_2 be Lie algebras over F. A homomorphism of Lie algebras is a linear map $\theta : g_1 \to g_2$ such that $\theta[xy] = [\theta x, \theta y]$ for all $x, y \in g_1$.

θ is an isomorphism of Lie algebras if θ is a bijective homomorphism.

Let g be a Lie algebra and h, k be subspaces of g. We define the product $[hk]$ to be the subspace spanned by all products $[xy]$ for $x \in h$, $y \in k$. Each element of $[hk]$ is thus a finite sum $x_1 y_1 + \cdots + x_r y_r$ with $x_i \in h$, $y_i \in k$. We note that $[hk] = [kh]$, i.e., multiplication of subspaces is commutative. This follows from the fact that multiplication of elements is anticommutative. So if $x \in h$, $y \in k$ we have $[yx] = -[xy] \in [hk]$.

A subalgebra of g is a subspace h of g such that $[hh] \subset h$.

An ideal of g is a subspace h of g such that $[hg] \subset h$.

We observe that, since $[hg] = [gh]$, there is no distinction in the theory of Lie algebras between left ideals and right ideals. Every ideal is two-sided.

Now let h be an ideal of the Lie algebra g. Let g/h be the vector space of cosets $h + x$ for $x \in g$. $h + x$ consists of all elements of form $y + x$ for $y \in h$. We claim that g/h can be made into a Lie algebra, the factor algebra of g with respect to h, by introducing the Lie multiplication

$$[h + x, h + y] = h + [xy].$$

We must take care to check that this operation is well defined, i.e., that if $h + x = h + x'$ and $h + y = h + y'$ then $h + [xy] = h + [x'y']$. This follows from the fact that h is an ideal of g. We have

$$x' = a + x, \quad y' = b + y \quad \text{for } a, b \in h.$$

Thus

$$[x'y'] = [ab] + [ay] + [xb] + [xy] \in h + [xy]$$

since $[ab]$, $[ay]$, $[xb]$ all lie in h. This gives $h+[x'y'] = h+[xy]$ as required.

There is a natural homomorphism $g \xrightarrow{\theta} g/h$ relating a Lie algebra with a factor algebra. θ is defined by $\theta(x) = h + x$. Conversely given any homomorphism $\theta : g_1 \to g_2$ of Lie algebras which is surjective, the kernel k of θ is an ideal of g_1 and the factor algebra g_1/k is isomorphic to g_2.

The set of all $n \times n$ matrices over the field F can be made into a Lie algebra under the Lie multiplication $[A, B] = AB - BA$. This Lie algebra is called $gl_n(F)$, the general linear Lie algebra of degree n over the field F.

1.2 Representations and modules

Let g be a Lie algebra over F. A representation of g is a homomorphism

$$\rho : g \to gl_n(F)$$

for some n. Two representations ρ, ρ' of g of degree n are called equivalent if there is a non-singular $n \times n$ matrix T over F such that

$$\rho'(x) = T^{-1}\rho(x)T, \quad \text{for all } x \in g.$$

There is a close connection between representations of g and g-modules. A left g-module is a vector space V over F with a multiplication

$$g \times V \to V$$
$$x, v \to xv$$

satisfying the axioms

(i) xv is linear in x and in v
(ii) $[xy]v = x(yv) - y(xv)$ for all $x, y \in g$, $v \in V$.

Every finite dimensional g-module gives a representation of g, as follows. Choose a basis $e_1, \ldots, e_n$ of V. Then xe_j is a linear combination of $e_1, \ldots, e_n$. Let

$$xe_j = \sum_{i=1}^{n} \rho_{ij}(x)e_i.$$

Let $\rho(x)$ be the $n \times n$ matrix $(\rho_{ij}(x))$. Then we have

$$\rho[xy] = \rho(x)\rho(y) - \rho(y)\rho(x) = [\rho(x)\,\rho(y)]$$

and so the map $x \to \rho(x)$ is a representation of g.

If we choose a different basis for the g-module V we shall get an equivalent representation.

Now let U be a subspace of V and h a subspace of g. Let hU be the subspace of V spanned by all elements xu for $x \in h$, $u \in U$. U is called a submodule of V if $gU \subset U$. A g-module V is called irreducible if V has no submodules other than V and 0.

Now g is itself a g-module under the multiplication $g \times g \to g$ given by $x, y \to [xy]$. To see this we must check $[[xy]z] = [x[yz]] - [y[xz]]$ for x, y, $z \in g$. This follows from the Jacobi identity using the anticommutative law. g is called the adjoint g-module, and it gives rise to the adjoint representation of g.

1.3 Special kinds of Lie algebra

So far the theory of Lie algebras has been very analogous to the theory of rings, where one has subrings, ideals, factor rings, etc. However there is also a sense in which the theory of Lie algebras can be considered as analogous to the theory of groups, where the Lie product $[xy]$ is regarded as analogous to the commutator $x^{-1}y^{-1}xy$ of two elements in a group. This analogy motivates the following terminology.

A Lie algebra g is called abelian if $[gg] = 0$. This means that all Lie products are zero.

We shall now define a sequence of subspaces $g^1, g^2, g^3, \cdots$ of g. We define them inductively by

$$g^1 = g, \quad g^{n+1} = [g^n g].$$

Now if h, k are ideals of g so is their product $[hk]$. For let $x \in h$, $y \in k$, $z \in g$. Then we have

$$[[xy]z] = [x[yz]] + [[xz]y] \in [hk].$$

Thus the product of two ideals is an ideal. It follows that all the subspaces g^i defined above are ideals of g. Thus we also have

$$g^{n+1} = [g^n g] \subset g^n$$

and so we have a descending series

$$g = g^1 \supset g^2 \supset g^3 \supset \cdots.$$

The Lie algebra g is called nilpotent if $g^i = 0$ for some i. Every abelian Lie algebra is nilpotent.

Example. The set of all $n \times n$ matrices (a_{ij}) over F with $a_{ij} = 0$ whenever $i \geq j$ is a nilpotent Lie algebra under Lie multiplication $[AB] = AB - BA$.

1.3 Special kinds of Lie algebra

We now define a different sequence of subspaces $g^{(0)}, g^{(1)}, g^{(2)}, \cdots$ of g. We again define them inductively by

$$g^{(0)} = g, \qquad g^{(n+1)} = [g^{(n)} g^{(n)}].$$

The $g^{(i)}$ are all ideals of g. Also we have

$$g^{(n+1)} = [g^{(n)} g^{(n)}] \subset g^{(n)}$$

and so we again have a descending series

$$g = g^{(0)} \supset g^{(1)} \supset g^{(2)} \supset \cdots.$$

The Lie algebra g is called soluble if $g^{(i)} = 0$ for some i.

Proposition. *Every nilpotent Lie algebra is soluble.*

Proof We show first that $[g^m g^n] \subset g^{m+n}$ for all m, n. We proceed by induction on n, the result being clear if $n = 1$. Assuming inductively that $[g^m g^n] \subset g^{m+n}$, let $x \in g^m$, $y \in g^n$, $z \in g$. Then we have

$$[x[yz]] = [[xy]z] - [[xz]y] \in g^{m+n+1}$$

by induction. Thus $[g^m g^{n+1}] \subset g^{m+n+1}$ as required.

We next observe that $g^{(n)} \subset g^{2^n}$. This is clear for $n = 0$. Assuming it inductively we have

$$g^{(n+1)} = [g^{(n)} g^{(n)}] \subset [g^{2^n} g^{2^n}] \subset g^{2^{n+1}}$$

as above. This completes the induction.

We now assume that g is nilpotent. Then $g^m = 0$ for some m. Hence there exists n with $g^{2^n} = 0$. It follows that $g^{(n)} = 0$ and so g is soluble. □

Example. The set of all $n \times n$ matrices (a_{ij}) over F with $a_{ij} = 0$ whenever $i > j$ is a soluble Lie algebra.

A Lie algebra g is called simple if g has no ideals other than g and 0.

A Lie algebra g of dimension 1 is of course simple because g has no proper subspaces at all. We have $g = Kx$ for some $x \in g$. Since $[xx] = 0$ we have $[gg] = 0$. Such a 1-dimensional Lie algebra will be called a trivial simple Lie algebra. We shall be mainly interested in non-trivial simple Lie algebras.

1.4 The Lie algebras $sl_n(\mathbb{C})$

We shall now take $F = \mathbb{C}$. Let $sl_n(\mathbb{C})$ be the set of all $n \times n$ matrices of trace 0. $sl_n(\mathbb{C})$ is an ideal of $gl_n(\mathbb{C})$. For if $A \in sl_n(\mathbb{C})$, $B \in gl_n(\mathbb{C})$ we have

$$\text{trace}[AB] = \text{trace}(AB - BA) = \text{trace}AB - \text{trace}BA = 0$$

since $\text{trace}AB = \text{trace}BA$ for any two $n \times n$ matrices. Hence $[AB] \in sl_n(\mathbb{C})$. Thus we see that $gl_n(\mathbb{C})$ is not simple.

$sl_n(\mathbb{C})$ is, however, a non-trivial simple Lie algebra when $n \geq 2$. To see this suppose we have a non-zero ideal k and take a non-zero element in this ideal. By multiplying this element on the left or right by suitable elementary matrices E_{ij} with $i \neq j$ we may simplify its form, while remaining within the ideal k. E_{ij} is the matrix with 1 in the i,j position and 0 elsewhere. Eventually we see that k contains some elementary matrix E_{ij}, and by further multiplication we see readily that k is the whole of $sl_n(\mathbb{C})$. Thus $sl_n(\mathbb{C})$ is simple.

We shall describle certain properties of $sl_n(\mathbb{C})$ in detail, because it is typical of simple Lie algebras in general.

Let h be the set of diagonal $n \times n$ matrices of trace 0. Then h is a subalgebra of $sl_n(\mathbb{C})$ and $\dim h = n - 1$. Furthermore we have $[hh] = 0$, so h is abelian.

We recall that g may be considered as a g-module, using $[gg] \subset g$. We thus have $[hg] \subset g$ and so we may regard g as a left h-module. We may write down a decomposition of g as a direct sum of h-submodules:

$$sl_n(\mathbb{C}) = h \oplus \sum_{i \neq j} \mathbb{C}E_{ij}.$$

We note that the 1-dimensional space $\mathbb{C}E_{ij}$ is an h-submodule since, for $x \in h$, we have

$$x = \begin{pmatrix} \lambda_1 & & 0 \\ & \ddots & \\ 0 & & \lambda_n \end{pmatrix}$$

with $\lambda_1 + \cdots + \lambda_n = 0$ and

$$[xE_{ij}] = (\lambda_i - \lambda_j)E_{ij}.$$

1.4 The Lie algebras $sl_n(\mathbb{C})$

This h-module gives a 1-dimensional representation of h

$$x = \begin{pmatrix} \lambda_1 & & \\ & \ddots & \\ & & \lambda_n \end{pmatrix} \longrightarrow \lambda_i - \lambda_j.$$

We note that there are $n(n-1)$ 1-dimensional representations of h arising in this way. They are called the roots of $sl_n(\mathbb{C})$ with respect to h. Let Φ be the set of roots. Φ lies in $h^* = \text{Hom}(h, \mathbb{C})$, the dual space of h.

We note that if $\alpha \in \Phi$ then $-\alpha \in \Phi$ also since the map $x \to \lambda_j - \lambda_i$ is the negative of the map $x \to \lambda_i - \lambda_j$. Thus the roots are certainly not linearly independent. The roots do however span h^*. For define $\alpha_i \in \Phi$ by

$$\alpha_i(x) = \lambda_i - \lambda_{i+1}.$$

Then $\alpha_1, \alpha_2, \ldots, \alpha_{n-1}$ are linearly independent and form a basis of h^*. Let $\Pi = \{\alpha_1, \alpha_2, \ldots, \alpha_{n-1}\}$. Π is called a set of fundamental roots, or simple roots. We consider the way in which the roots are expressed as linear combinations of the fundamental roots. The root $x \to \lambda_i - \lambda_j$ is equal to

$$\alpha_i + \alpha_{i+1} + \ldots + \alpha_{j-1} \quad \text{if } i < j$$

and to

$$-(\alpha_j + \alpha_{j+1} + \ldots + \alpha_{i-1}) \quad \text{if } i > j.$$

Thus each root in Φ is a linear combination of fundamental roots with coefficients in $\mathbb{Z}$ which are either all non-negative or all non-positive. Thus we may write $\Phi = \Phi^+ \cup \Phi^-$ where Φ^+ consists of positive combinations of Π and Φ^- negative combinations.

We shall keep this example $sl_n(\mathbb{C})$ in mind to illustrate the general theory of simple Lie algebras.

2
Simple Lie algebras over $\mathbb{C}$

2.1 Cartan subalgebras

Let g be a finite dimensional Lie algebra over $\mathbb{C}$. For any subalgebra h of g we define $I(h)$ by

$$I(h) = \{x \in g; [yx] \in h \text{ for all } y \in h\}.$$

It is readily checked that $I(h)$ is a subalgebra of g containing h, and that h is an ideal of $I(h)$. Moreover if h is an ideal of k then k is contained in $I(h)$. Thus $I(h)$ is the largest subalgebra of g in which h is an ideal. $I(h)$ is called the idealizer of h.

Definition A subalgebra h of g is called a Cartan subalgebra if h is nilpotent and $h = I(h)$.

Theorem *Every finite dimensional Lie algebra g over $\mathbb{C}$ has a Cartan subalgebra. Moreover given any two Cartan subalgebras h_1, h_2 of g there exists an automorphism θ of g (i.e. an isomorphism of g into itself) such that $\theta(h_1) = h_2$.*

Proof We shall not give the proof of this theorem, which is lengthy, but shall indicate briefly how a Cartan subalgebra can be obtained. For any element $x \in g$ we define the linear map $\text{ad} x : g \to g$ by

$$\text{ad} x.y = [xy]$$

Given any $\lambda \in \mathbb{C}$ the eigenspace of $\text{ad} x$ with eigenvalue λ is

$$\{y \in g; (\text{ad} x - \lambda 1)y = 0\}.$$

The generalized eigenspace of $\text{ad} x$ with eigenvalue λ is

$$\{y \in g; (\text{ad} x - \lambda 1)^i y = 0 \text{ for some } i\}.$$

It is well known from linear algebra that g is the direct sum of its generalized eigenspaces for all $\lambda \in \mathbb{C}$. (This is not true for the ordinary eigenspaces unless $\operatorname{ad} x$ can be represented by a diagonal matrix). Let $h(x)$ be the generalized eigenspace of g with respect to $\operatorname{ad} x$ with eigenvalue $\lambda = 0$. We say that x is a regular element of g if the dimension of $h(x)$ is as small as possible. It turns out that when x is regular $h(x)$ is a Cartan subalgebra of g. $\square$

The fact that any two Cartan subalgebras are related by some automorphism of g is proved using a density argument and ideas from algebraic geometry. In fact one can use a special kind of automorphism of g, called an inner automorphism, to transform h_1 to h_2.

Example Let $g = sl_n(\mathbb{C})$ and h be the subalgebra of diagonal matrices in g. Then h is a Cartan subalgebra of g.

Since $[hh] = 0$, h is clearly nilpotent. To show $h = I(h)$ let $\sum_{i,j} a_{ij} E_{ij}$ be any element of $I(h)$. Choose $p, q \in \{1, \cdots, n\}$ with $p \neq q$. Then $E_{pp} - E_{qq} \in h$, hence

$$\left[\sum_{i,j} a_{ij} E_{ij}, E_{pp} - E_{qq} \right] \in h.$$

This gives

$$\sum_i a_{ip} E_{ip} - \sum_i a_{iq} E_{iq} - \sum_j a_{pj} E_{pj} - \sum_j a_{qj} E_{qj} \in h.$$

Since this matrix is diagonal we deduce, by considering the coefficient of E_{pq}, that $a_{pq} = 0$. Since this is true for all p, q with $p \neq q$ we have $\sum a_{ij} E_{ij} \in h$. Thus $h = I(h)$.

It turns out in fact that whenever g is a simple Lie algebra its Cartan subalgebras are abelian, as is the case in $sl_n(\mathbb{C})$.

2.2 The Cartan decomposition

Let g be a simple non-trivial Lie algebra over $\mathbb{C}$ and h be a Cartan subalgebra of g. Then we have $[hh] = 0$. Since we have $[hg] \subset g$ we make regard g as a left h-module. h is then a submodule of g. In fact it is possible to express g as the direct sum of h with a number of 1-dimensional h-submodules. Such a decomposition is uniquely determined by h. It is called the Cartan decomposition of g with respect to h. We write it as

$$g = h \oplus \sum_\alpha \mathbb{C} e_\alpha$$

$\mathbb{C} e_\alpha$ is a 1-dimensional h-module, thus we have

$$[x e_\alpha] = \alpha(x) e_\alpha \quad \alpha(x) \in \mathbb{C}$$

for all $x \in h$. α lies in the dual space $h^* = \text{Hom}(h, \mathbb{C})$ of h. The 1-dimensional representations α of h arising in the Cartan decomposition are called the roots of g with respect to h. The set of roots will be denoted by Φ. Thus we have

$$g = h \oplus \sum_{\alpha \in \Phi} \mathbb{C} e_\alpha$$

and $\dim g = \dim h + |\Phi|$. The Cartan decomposition in the case $g = sl_n(\mathbb{C})$ was described in detail in §1.4.

The root system Φ has the following properties. If $\alpha \in \Phi$ then $-\alpha \in \Phi$. Also the set Φ spans h^*. However Φ is not linearly independent, so it is natural to choose a subset of Φ which will form a basis for h^*. In fact such a basis can be chosen in rather a special way. There exists a subset Π of Φ, called the set of fundamental roots, such that Π is linearly independent and each $\alpha \in \Phi$ can be expressed as a linear combination of roots in Π with coefficients in $\mathbb{Z}$ which are either all $\geqslant 0$ or all $\leqslant 0$. Such a system Π was given when $g = sl_n(\mathbb{C})$ in §1.4. The choice of the system Π of fundamental roots is not unique. However once Π is chosen we can define the sets Φ^+ and Φ^- of positive and negative roots. We have

$$\Phi = \Phi^+ \cup \Phi^- \text{ and } \Phi^- = -\Phi^+.$$

We shall denote by $h_\mathbb{R}^*$ the set of elements of h^* which are linear combinations of elements of Π with coefficients in $\mathbb{R}$. The definition of $h_\mathbb{R}^*$ is in fact independent of the choice of Π since it consists of all real combinations of elements of Φ. We have

$$\dim_\mathbb{R} h_\mathbb{R}^* = \dim_\mathbb{C} h^* = \dim_\mathbb{C} h = l.$$

l is called the rank of the Lie algebra g.

2.3 The Killing form

We consider the map
$$g \times g \to \mathbb{C} \quad \text{defined by} \quad <x,y> = \text{trace}(\text{ad}x\,\text{ad}y).$$
$$x, y \to <x, y>$$
$\text{ad}x$, $\text{ad}y$ and $\text{ad}x\,\text{ad}y$ are linear maps of g into itself. $\text{trace}(\text{ad}x\,\text{ad}y)$ is the trace of any matrix representing $\text{ad}x\,\text{ad}y$, and is independent of the choice of such a matrix. Since $\text{trace}AB = \text{trace}BA$ for any two square matrices A, B we have $<x,y> = <y,x>$. Thus we have a symmetric bilinear form on g. This is called the *Killing form*.

We now assume that g is a non-trivial simple Lie algebra over $\mathbb{C}$. Then the Killing form on g is non-degenerate in the sense that

$$<x,y> = 0 \text{ for all } y \in g \text{ implies } x = 0.$$

We may restrict the Killing form on g to h, to give a map $h \times h \to \mathbb{C}$. It can be shown that this map remains non-degenerate on h. Thus

$$x \in h \text{ and } <x,y> = 0 \text{ for all } y \in h \text{ implies } x = 0.$$

We may thus define a map $h \to h^*$ given by $x \to f_x$ where

$$f_x(y) = <x, y> \text{ for all } y \in h.$$

This is a linear map from h to h^*. Since the Killing form is non-degenerate on h this map is bijective. Thus each element of h^* has form f_x for just one $x \in h$. We may thus define a map $h^* \times h^* \to \mathbb{C}$ by

$$<f_x, f_y> = <x, y> \text{ for } x, y \in h.$$

We may restrict this bilinear form to the real vector space $h_{\mathbb{R}}^*$. It can be shown that its values then lie in $\mathbb{R}$. Thus we have a map

$$h_{\mathbb{R}}^* \times h_{\mathbb{R}}^* \to \mathbb{R}.$$

This map has the property that

$$<\lambda, \lambda> \geqslant 0 \text{ for all } \lambda \in h_{\mathbb{R}}^*.$$

Moreover $<\lambda, \lambda> = 0$ implies $\lambda = 0$. Thus the scalar product on $h_{\mathbb{R}}^*$ is positive definite. $h_{\mathbb{R}}^*$ is therefore a Euclidean space.

This Euclidean space $h_{\mathbb{R}}^*$ contains the set of roots Φ. The properties of the configuration formed by the roots in $h_{\mathbb{R}}^*$ is important in the classification of the simple Lie algebras g.

16   I Lie Algebras

Examples. Let $g = sl_2(\mathbb{C})$. Then $\dim h = 1$. Let $\Pi = \{\alpha_1\}$. Then $\Phi = \{\alpha_1, -\alpha_1\}$. The configuration formed by Φ in the 1-dimensional Euclidean space $h_\mathbb{R}^*$ is

$$\underline{\qquad\qquad\qquad\quad}_{-\alpha_1}\quad\underline{\qquad 0 \qquad}\quad_{\alpha_1}\underline{\qquad\qquad\qquad\quad}$$

Now let $g = sl_3(\mathbb{C})$. Then $\dim h = 2$. Let $\Pi = \{\alpha_1, \alpha_2\}$. Then, as shown in §1.4, we have $\Phi = \{\alpha_1, \alpha_2, \alpha_1 + \alpha_2, -\alpha_1, -\alpha_2, -\alpha_1 - \alpha_2\}$.

The configuration formed by Φ in the 2-dimension Euclidean space $h_\mathbb{R}^*$ is

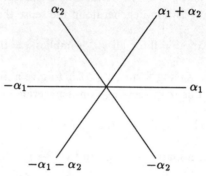

2.4 The Weyl group

The configuration formed by the root system Φ is best understood by introducing a certain group of non-singular linear transformations of $h_\mathbb{R}^*$ called the Weyl group. For each $\alpha \in \Phi$ let $s_\alpha : h_\mathbb{R}^* \to h_\mathbb{R}^*$ be the map defined by

$$s_\alpha(\lambda) = \lambda - 2\frac{<\alpha, \lambda>}{<\alpha, \alpha>}\alpha.$$

Note that $s_\alpha(\alpha) = -\alpha$ and $s_\alpha(\lambda) = \lambda$ whenever $<\alpha, \lambda> = 0$. Thus s_α is the reflection in the hyperplane orthogonal to α. Let W be the group generated by the maps s_α for all $\alpha \in \Phi$. W is called the Weyl group.

W has some favourable properties. In the first place it permutes the roots, i.e. $w(\alpha) \in \Phi$ for all $\alpha \in \Phi$ and all $w \in W$. It follows that W is finite, since there are only finitely many permutations of Φ, and each such permutation comes from at most one linear transformation since Φ spans $h_\mathbb{R}^*$. Also we have $\Phi = W(\Pi)$, i.e., given any $\alpha \in \Phi$ there exists $\alpha_i \in \Pi$ and $w \in W$ such that $\alpha = w(\alpha_i)$. Moreover W is generated by the s_{α_i} for $\alpha_i \in \Pi$.

The importance of the Weyl group is that it enables us to reconstruct

2.4 The Weyl group

the full root system Φ given only the set Π of fundamental roots. For given Π the Weyl group is determined, being the group generated by the reflections s_{α_i} for $\alpha_i \in \Pi$. The root system Φ is then determined, since $\Phi = W(\Pi)$. Hence, given Π, the root system Φ is obtained by successive reflections s_{α_i} until no further vectors can be obtained.

An example when $g = sl_3(\mathbb{C})$ is shown in the figure.

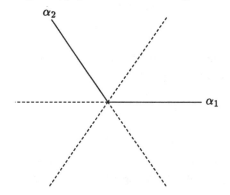

Given α_1, α_2 the remaining roots are obtained by reflecting successively by $s_{\alpha_1}, s_{\alpha_2}$.

We note that
$$s_{\alpha_i}(\alpha_j) = \alpha_j - \frac{2 <\alpha_i, \alpha_j>}{<\alpha_i, \alpha_i>} \alpha_i.$$

If $\alpha_i, \alpha_j \in \Pi$ with $i \neq j$ $s_{\alpha_i}(\alpha_j)$ is a root, so is a $\mathbb{Z}$-combination of α_i and α_j. Since the coefficient of α_j is 1 the coefficient of α_i must be a non-negative integer, since the given root lies in Φ^+. It follows that

$$\frac{2 <\alpha_i, \alpha_j>}{<\alpha_i, \alpha_i>} \in \mathbb{Z}, \quad \frac{2 <\alpha_i, \alpha_j>}{<\alpha_i, \alpha_i>} \leq 0.$$

We define $A_{ij} = \frac{2 <\alpha_i, \alpha_j>}{<\alpha_i, \alpha_i>}$.

The numbers A_{ij} are called the Cartan integers and the matrix $A = (A_{ij})$ which they form is called the Cartan matrix. We have $A_{ij} \in \mathbb{Z}$, $A_{ii} = 2$ and $A_{ij} \leq 0$ if $i \neq j$.

Let θ_{ij} be the angle between α_i, α_j. This angle can be determined by the cosine formula

$$<\alpha_i, \alpha_j> = <\alpha_i, \alpha_i>^{1/2} <\alpha_j, \alpha_j>^{1/2} \cos \theta_{ij}.$$

Thus we have
$$4 \cos^2 \theta_{ij} = \frac{2 <\alpha_i, \alpha_j>}{<\alpha_i, \alpha_i>} \cdot \frac{2 <\alpha_j, \alpha_i>}{<\alpha_j, \alpha_j>}.$$

Hence $4 \cos^2 \theta_{ij} = A_{ij} A_{ji}$.

We shall write $n_{ij} = A_{ij}A_{ji}$. Then $n_{ij} \in \mathbb{Z}$ and $n_{ij} \geq 0$. Moreover, since

$$-1 \leq \cos \theta_{ij} \leq 1$$

we have

$$0 \leq 4\cos^2 \theta_{ij} \leq 4.$$

In fact when $i \neq j$ we have $\theta_{ij} \neq 0$ and so

$$0 \leq 4\cos^2 \theta_{ij} < 4.$$

Hence the only possible values for n_{ij} are $n_{ij} = 0, 1, 2, 3$.

We shall now encode this information about the system Π of fundamental roots in terms of a graph.

2.5 The Dynkin diagram

The Dynkin diagram Δ of g is the graph with nodes labelled $1, \cdots, l$ in bijective correspondence with the set Π of fundamental roots, such that nodes i, j with $i \neq j$ are joined by n_{ij} bonds.

Example Let $g = sl_3(\mathbb{C})$. Then $\Pi = \{\alpha_1, \alpha_2\}$ and $s_{\alpha_1}(\alpha_2) = \alpha_1 + \alpha_2$, $s_{\alpha_2}(\alpha_1) = \alpha_1 + \alpha_2$.

Thus $A_{12} = -1$, $A_{21} = -1$ and so $n_{12} = 1$. Thus Δ is the graph

$$\underset{1}{\circ} \!\!\!-\!\!\!-\!\!\!-\!\!\!-\!\!\!-\!\!\!-\!\!\! \underset{2}{\circ}$$

The Dynkin diagram is uniquely determined by g. The choice of Cartan subalgebra does not matter since any two Cartan subalgebras are related by some automorphism of g. The choice of fundamental system Π does not matter, since it can be shown that any two fundamental systems Π_1, Π_2 have the property that $\Pi_2 = w(\Pi_1)$ for some $w \in W$.

The Dynkin diagram of g has the following properties. Δ is a connected graph provided g is a non-trivial simple Lie algebra. Any two nodes are joined by at most 3 bonds. Also let $Q(x_1, \cdots, x_l)$ be the quadratic form

$$Q(x_1, \cdots, x_l) = 2\sum_{i=1}^{l} x_i^2 - \sum_{\substack{i,j \\ i \neq j}} \sqrt{n_{ij}} x_i x_j.$$

This quadratic form is determined by the Dynkin diagram. For example if Δ is $\underset{1}{\circ}\!\!-\!\!\underset{2}{\circ}$ then we have

$$Q(x_1, x_2) = 2x_1^2 + 2x_2^2 - 2x_1 x_2.$$

2.5 The Dynkin diagram

Now the quadratic form $Q(x_1, \cdots, x_l)$ is positive definite, since we have

$$Q(x_1, \cdots, x_l) = 2 \left\langle \frac{\sum x_i \alpha_i}{\sqrt{<\alpha_i, \alpha_i>}}, \frac{\sum x_i \alpha_i}{\sqrt{<\alpha_i, \alpha_i>}} \right\rangle.$$

We shall consider the problem of determining all graphs Δ with the above properties.

Theorem *Consider graphs Δ with the following properties:*

(a) Δ *is connected.*

(b) *The number of bonds joining any two nodes is* $0, 1, 2, 3$.

(c) *The quadratic form Q determined by Δ is positive definite.*

Then Δ must be one of the graphs on the following list,

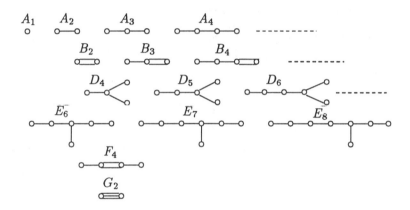

Fig. 2.1. list 1.

The graphs on this list will be called Dynkin diagrams.

Proof A subgraph of a graph Δ is one obtained from Δ by removing certain nodes or decreasing certain bond strengths or both. For example ⊂⊃ is a subgraph of ○—⊂⊃ . The list of graphs given in the theorem will be called the standard list. We note that each subgraph of a graph on the standard list is also on the standard list. It is not difficult to show that if the quadratic form of a graph Δ is positive definite, then the quadratic form of any subgraph of Δ is positive definite also.

I Lie Algebras

Now the quadratic form of a graph Δ is represented by a symmetric matrix

$$M = \begin{pmatrix} 2 & & \\ & \ddots & -n_{ij} \\ & & 2 \end{pmatrix}.$$

We recall from linear algebra that $Q(x_1, \cdots, x_l)$ is positive definite if and only if all the leading minors of M have positive determinant. However the leading minors of M are simply matrices M corresponding to certain subgraphs of Δ. In order to show that $Q(x_1, \cdots, x_l)$ is positive definite it is therefore sufficient to check that $\det M > 0$ for each graph Δ on the standard list. This is readily verified.

We now wish to prove conversely that the graphs on the standard list are the only ones satisfying the given conditions. In order to do this we introduce a second list.

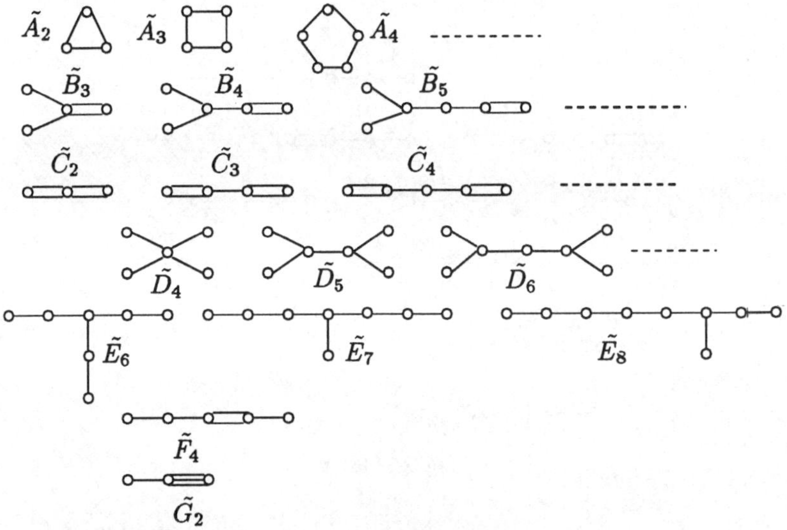

Fig. 2.2. list 2.

It may be readily checked that each graph Δ on list 2 has a quadratic form $Q(x_1, \cdots, x_l)$ with symmetric matrix M satisfying $\det M = 0$. Thus $Q(x_1, \cdots, x_l)$ is not positive definite. Hence any graph Δ satisfying our given conditions can contain no subgraph on list 2.

Let Δ be a graph satisfying our conditions (a), (b), (c). Then Δ has no cycles, otherwise Δ would contain a subgraph of type $\tilde{A}_l$. Δ has at

2.5 The Dynkin diagram

most one multiple bond, otherwise Δ would contain a subgraph of type $\widetilde{C}_l$. Δ cannot have both a multiple bond and a branch point, otherwise Δ would contain a subgraph $\widetilde{B}_l$. Also Δ cannot have more than one branch point, otherwise Δ would contain a subgraph $\widetilde{D}_l$.

Suppose Δ has a triple bond. Then Δ must be G_2, as otherwise Δ would contain a subgraph $\widetilde{G}_2$. We may therefore assume that Δ contains no triple bond.

Suppose Δ has a double bond. Then Δ contains no branch point, so is a chain. If the double bond is at one end of the chain then $\Delta = B_l$. If not Δ must be F_4, since otherwise Δ would contain a subgraph $\widetilde{F}_4$.

Thus we may assume Δ contains only single bonds. If Δ has no branch point then $\Delta = A_l$. Thus we suppose that Δ contains a branch point. This branch point has only 3 branches, otherwise Δ would contain a subgraph $\widetilde{D}_4$. Let the lengths of the branches be l_1, l_2, l_3 with $l = l_1 + l_2 + l_3 + 1$ and $l_1 \geqslant l_2 \geqslant l_3$. Then $l_3 = 1$, otherwise Δ would contain a subgraph $\widetilde{E}_6$. Also $l_2 \leqslant 2$ otherwise Δ would contain a subgraph $\widetilde{E}_7$. If $l_2 = 1$ then $\Delta = D_l$. So we may suppose $l_2 = 2$. We then have $l_1 \leqslant 4$, otherwise Δ would contain a subgraph $\widetilde{E}_8$. If $l_1 = 2$ then $\Delta = E_6$. If $l_1 = 3$ then $\Delta = E_7$. If $l_1 = 4$ then $\Delta = E_8$.

Thus Δ must be one of the graphs on the standard list. $\square$

We now consider to what extent the Dynkin diagram determines the matrix of Cartan integers. We recall that

$$n_{ij} = A_{ij} A_{ji} \qquad i \neq j$$

and that A_{ij}, A_{ji} are integers $\leqslant 0$. Moreover $A_{ij} = 0$ if and only if $A_{ji} = 0$.

If $n_{ij} = 0$ we must therefore have $A_{ij} = 0$ and $A_{ji} = 0$. If $n_{ij} = 1$ we must have $A_{ij} = -1$ and $A_{ji} = -1$. If $n_{ij} = 2$ however, there are two possible factorisations of n_{ij}. Either we have $A_{ij} = -1, A_{ji} = -2$ or we have $A_{ij} = -2, A_{ji} = -1$. Since

$$A_{ij} = \frac{2 <\alpha_i, \alpha_j>}{<\alpha_i, \alpha_i>}$$

we have

$$\frac{A_{ij}}{A_{ji}} = \frac{<\alpha_j, \alpha_j>}{<\alpha_i, \alpha_i>}.$$

Thus in the first case above we have $<\alpha_i, \alpha_i> > <\alpha_j, \alpha_j>$ and in the second case $<\alpha_i, \alpha_i> < <\alpha_j, \alpha_j>$. We distinguish between these two cases by putting an arrow on the Dynkin diagram pointing towards the long root. (The arrow can be interpreted as an inequality of root lengths.) In the first case we have the diagram

$$i \;\text{O}\!\!\Longrightarrow\!\!\text{O}\; j$$

and in the second case the diagram

$$i \;\text{O}\!\!\Longleftarrow\!\!\text{O}\; j$$

Similarly if $n_{ij} = 3$ we get two possible factorisations $n_{ij} = A_{ij}A_{ji}$ which are distinguished by putting an arrow on the given triple bond.

In the cases when Δ is B_2, F_4, G_2, it does not matter in which direction the arrow is inserted, since the graphs are symmetric. However when Δ is B_l for $l \geqslant 3$ we can obtain two different diagrams by inserting an arrow. Those diagrams will be labelled B_l, C_l as shown:

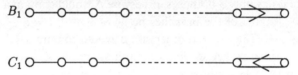

Thus in type B_l the last fundamental root is shorter than the others, whereas in type C_l it is longer than the others.

The main theorem on the classification of the finite dimensional simple Lie algebras over $\mathbb{C}$ is as follows.

Theorem. *Let g be a finite dimensional simple non-trivial Lie algebra over $\mathbb{C}$. Then the Cartan matrix of g is one of those on the standard list*

$$A_l \;\; l \geqslant 1, \qquad B_l \;\; l \geqslant 2, \qquad C_l \;\; l \geqslant 3, \qquad D_l \;\; l \geqslant 4,$$
$$E_6, \quad E_7, \quad E_8, \quad F_4, \quad G_2.$$

Moreover for any Cartan matrix on the standard list there is just one simple Lie algebra, up to isomorphism, giving rise to it.

The classification of the simple Lie algebras was achieved by W. Killing, in a series of papers in Mathematische Annalen between 1888 and 1890, and independently by E. Cartan in his thesis in Paris in 1894.

The dimensions of the simple Lie algebras may be calculated as follows. The Dynkin diagram determines the configuration formed by the set Π of fundamental roots, i.e., the angles between the fundamental roots and their relative lengths. We may then obtain the full root system Φ by successive reflection by elements of the Weyl group, as explained earlier. Finally, since we have

$$g = h \oplus \sum_{\alpha \in \Phi} \mathbb{C}e_\alpha$$

2.5 The Dynkin diagram

it follows that

$$\dim g = \dim h + |\Phi|.$$

The dimensions of the simple Lie algebras are given in the following table.

$$\dim A_l = l(l+2)$$
$$\dim B_l = l(2l+1)$$
$$\dim C_l = l(2l+1)$$
$$\dim D_l = l(2l-1)$$
$$\dim G_2 = 14$$
$$\dim F_4 = 52$$
$$\dim E_6 = 78$$
$$\dim E_7 = 133$$
$$\dim E_8 = 248.$$

The algebras of classical type A_l, B_l, C_l, D_l can be described conveniently in terms of matrices. The simple Lie algebra A_l is isomorphic to the Lie algebra $sl_{l+1}(\mathbb{C})$ of all $(l+1) \times (l+1)$ matrices of trace 0. The simple Lie algebra D_l is isomorphic to the Lie algebra $so_{2l}(\mathbb{C})$ of all $2l \times 2l$ skew-symmetric matrices. Although this is the simplest description of this Lie algebra, another is more convenient. D_l is isomophic to the Lie algebra of all $2l \times 2l$ matrices T satisfying the condition

$$TA + AT^t = 0$$

where $A = \begin{pmatrix} 0 & I_l \\ I_l & 0 \end{pmatrix}$. The advantage of this description of g is that the diagonal matrices in g form a Cartan subalgebra h, and the Cartan decomposition can be readily obtained.

The simple Lie algebra B_l is isomorphic to the Lie algebra $so_{2l+1}(\mathbb{C})$ of all $(2l+1) \times (2l+1)$ skew-symmetric matrices T. It is also isomorphic to the Lie algebra of all $(2l+1) \times (2l+1)$ matrices satisfying $TA + AT^t = 0$ where

$$A = \begin{pmatrix} 1 & 0 & \cdots & 0 \\ 0 & 0 & & I_l \\ \vdots & & & \\ 0 & I_l & & 0 \end{pmatrix}$$

The simple Lie algebra C_l is isomorphic to the Lie algebra of all $2l \times 2l$

matrices T satisfying $TA + AT^t = 0$ where

$$A = \begin{pmatrix} 0 & I_1 \\ -I_1 & 0 \end{pmatrix}.$$

Each of these Lie algebras is the Lie algebra Lie G of some Lie group G, i.e., the tangent space of G at the identity element with suitable Lie multiplication. (G is not uniquely determined up to isomorphism by its Lie algebra). For the classical types we have

$$\begin{aligned} sl_{l+1}(\mathbb{C}) &= \text{Lie } SL_{l+1}(\mathbb{C}) & \text{Type } A_l \\ so_{2l+1}(\mathbb{C}) &= \text{Lie } SO_{2l+1}(\mathbb{C}) & \text{Type } B_l \\ sp_{2l}(\mathbb{C}) &= \text{Lie } Sp_{2l}(\mathbb{C}) & \text{Type } C_l \\ so_{2l}(\mathbb{C}) &= \text{Lie } SO_{2l}(\mathbb{C}) & \text{Type } D_l \end{aligned}$$

The exceptional simple Lie algebras G_2, F_4, E_6, E_7, E_8 can be constructed in terms of the Cayley algebra or algebra of octonians σ. Given any algebra A a derivation $D : A \mapsto A$ is a linear map such that

$$D(ab) = Da \cdot b + a \cdot Db.$$

If D_1, D_2 are derivations of A so is $[D_1\ D_2] = D_1 D_2 - D_2 D_1$. The derivations of A form a Lie algebra Der$A$.

The Lie algebra of derivations of the Cayley algebra σ over $\mathbb{C}$ is the simple Lie algebra G_2.

The vector space of all 3×3 hermitian matrices over the Cayley algebra σ forms a Jordan algebra J under the operation

$$A \cdot B = \frac{AB + BA}{2}.$$

We have dim $J = 27$. The Lie algebra of all derivations of J is the simple Lie algebra F_4. The algebras E_6, E_7, E_8 can also be constructed by making use of σ and J in different ways.

3
Representations of simple Lie algebras

In the present section we shall discuss the finite dimensional irreducible g-modules, where g is a simple non-trivial Lie algebra.

3.1 The universal enveloping algebra

Let g be any finite dimensional Lie algebra over $\mathbb{C}$. Let $T(g)$ be the tensor algebra of g.

$$T(g) = \mathbb{C}1 \oplus g \oplus (g \otimes g) \oplus (g \otimes g \otimes g) \oplus \cdots$$

$T(g)$ is a vector space over $\mathbb{C}$ on which a multiplication is defined in a natural way. Let I be the 2-sided ideal of $T(g)$ generated by all elements of the form

$$x \otimes y - y \otimes x - [xy] \text{ for } x, y \in g.$$

Let $U(g) = T(g)/I$. $U(g)$ is an associative algebra called the universal enveloping algebra of g.

A basis of $U(g)$ can be obtained as follows. If $x_1, \cdots, x_n$ is a basis of g then the set of elements

$$x_1^{i_1} x_2^{i_2} \cdots x_n^{i_n} \qquad i_r \in \mathbb{Z}, i_r \geqslant 0$$

forms a basis of $U(g)$. This is called the Poincaré-Birkhoff-Witt basis theorem. In the special case when $[gg] = 0$, i.e., g is abelian, we have $x_i x_j = x_j x_i$ in $U(g)$ and so $U(g)$ is isomorphic to the polynomial ring $\mathbb{C}[x_1, \ldots x_n]$. In general, however, x_i, x_j do not commute and we have instead

$$x_i x_j - x_j x_i = [x_i x_j].$$

Thus $U(g)$ is a kind of non-commutative polynomial ring.

The importance of the enveloping algebra $U(g)$ is that it has the same representation theory as g. If V is a g-module then V can be regarded as a $T(g)$-module in a natural way. Since

$$[xy]v = x(yv) - y(xv)$$

for $x, y \in g$, $v \in V$ we see that

$$(x \otimes y - y \otimes x - [xy])v = 0$$

for all $v \in V$. Thus elements $x \otimes y - y \otimes x - [xy]$ lie in the kernel of V. This kernel is a 2-sided ideal of $T(g)$, so contains I. Thus V may be regarded as a $U(g)$-module since $U(g) = T(g)/I$.

Conversely every $U(g)$-module may be regarded as a g-module using the map

$$g \to T(g) \twoheadrightarrow U(g).$$

This map is injective by the PBW-basis theorem, and so g may be regarded as a subspace of $U(g)$.

3.2 Verma modules

We now suppose that g is a non-trivial simple Lie algebra. Let h be a Cartan subalgebra of g and

$$g = h \oplus \sum_{\alpha \in \Phi} \mathbb{C} e_\alpha$$

be the Cartan decomposition of g with respect to h. We recall that the Killing form gives a bijection $h \leftrightarrow h^*$. Let $h_i \in h$ be the element corresponding to $\frac{2\alpha_i}{<\alpha_i, \alpha_i>} \in h^*$ under this bijection. Then

$$\alpha_j(h_i) = \frac{2 <\alpha_i, \alpha_j>}{<\alpha_i, \alpha_i>} = A_{ij} \in \mathbb{Z}.$$

Thus all the fundamental roots $\alpha_1, \cdots, \alpha_l$ take integer values at h_i. $h_1, \cdots, h_l$ form a basis of h. They are called the fundamental coroots.

Let $\lambda \in h^*$ and $J(\lambda)$ be the left ideal of $U(g)$ generated by the elements e_α, $\alpha \in \Phi^+$, and $h_i - \lambda(h_i)1$ for $i = 1, \cdots, l$. Thus

$$J(\lambda) = \sum_{\alpha \in \Phi^+} U(g)e_\alpha + \sum_{i=1}^{l} U(g)(h_i - \lambda(h_i)1).$$

$J(\lambda)$ is a $U(g)$-submodule of $U(g)$. Let $M(\lambda) = U(g)/J(\lambda)$. Then $M(\lambda)$ is

3.3 Finite dimensional irreducible modules

also a $U(g)$-module, called the Verma module determined by λ. We have a natural homomorphism

$$U(g) \xrightarrow{\theta} M(\lambda)$$

of left $U(g)$-modules. Let $m_\lambda = \theta(1)$. Then we have

$$e_\alpha m_\lambda = 0 \text{ for all } \alpha \in \Phi^+$$
$$h_i m_\lambda = \lambda(h_i) m_\lambda \text{ for } i = 1, \cdots, l.$$

Since each element $u \in U(g)$ satisfies $u = u1$, each element of $M(\lambda)$ has the form $u m_\lambda$ for some $u \in U(g)$. Thus

$$M(\lambda) = U(g) m_\lambda$$

is a cyclic $U(g)$-module.

We may regard $M(\lambda)$ as an h-module. $M(\lambda)$ decomposes into the direct sum of 1-dimensional h-submodules. The 1-dimensional representations $\mu \in h^*$ obtained form these submodules are called the weights of $M(\lambda)$. λ is a weight of $M(\lambda)$ since

$$h_i m_\lambda = \lambda(h_i) m_\lambda.$$

All the weights of $M(\lambda)$ turn out to have the form

$$\lambda - m_1 \alpha_1 - \cdots - m_l \alpha_l$$

where $\alpha_1, \cdots, \alpha_l$ are the fundamental roots and $m_1, \cdots, m_l$ are non-negative integers. λ is thus in a natural sense the highest weight of $M(\lambda)$. $M(\lambda)$ is called the Verma module with highest weight λ.

It can be shown that $M(\lambda)$ has a unique maximal submodule $K(\lambda)$. Let $L(\lambda) = M(\lambda)/K(\lambda)$. Then $L(\lambda)$ is an irreducible $U(g)$-module.

We thus have a procedure for constructing irreducible g-modules. For each $\lambda \in h^*$ we have obtained an irreducible g-module $L(\lambda)$ as the top quotient of the Verma module $M(\lambda)$. $L(\lambda)$ is not necessarily finite dimensional – this depends on the choice of λ.

3.3 Finite dimensional irreducible modules

Theorem $\dim L(\lambda)$ *is finite if and only if* $\lambda(h_i) \in \mathbb{Z}$, $\lambda(h_i) \geqslant 0$ *for all* $i = 1, \cdots, l$.

$\lambda \in h^*$ is called integral if $\lambda(h_i) \in \mathbb{Z}$ for all i and dominant integral if in addition $\lambda(h_i) \geqslant 0$ for all i.

Theorem *Every finite dimensional irreducible g-module has the form $L(\lambda)$ for some dominant integral $\lambda \in h^*$.*

Thus we have a bijective correspondence between finite dimensional irreducible g-modules, up to isomorphism, and dominant integral weights λ.

This classification of the irreducible g-modules goes back to E. Cartan's thesis of 1894.

The dominant integral weights can be described conveniently in the following way. Let $\omega_i \in h^*$ satisfy

$$\omega_i(h_i) = 1$$
$$\omega_i(h_j) = 0 \text{ for } j \neq i.$$

The elements $\omega_1, \cdots, \omega_l$ of h^* uniquely determined in this way are called the fundamental weights. They form a basis of h^*. Let $\lambda \in h^*$ and

$$\lambda = m_1\omega_1 + \cdots + m_l\omega_l.$$

Then $\lambda(h_i) = m_i\omega_i(h_i) = m_i$. Thus we have

$$\lambda = \lambda(h_1)\omega_1 + \cdots + \lambda(h_l)\omega_l.$$

We see from this that the dominant integral weights are precisely the non-negative integral combinations of the fundamental weights.

We consider the relation between the fundamental weights $\omega_1, \cdots, \omega_l$ and the fundamental roots $\alpha_1, \cdots, \alpha_l$. Let

$$\alpha_i = \sum_{j=1}^{l} m_{ij}\omega_j.$$

Then we have $\alpha_i(h_j) = m_{ij}\omega_j(h_j) = m_{ij}$. By definition of h_j we have

$$m_{ij} = \alpha_i(h_j) = \left\langle \alpha_i, \frac{2\alpha_j}{<\alpha_j, \alpha_j>} \right\rangle = \frac{2 <\alpha_j, \alpha_i>}{<\alpha_j, \alpha_j>} = A_{ji}.$$

Thus

$$\alpha_i = \sum_{j=1}^{l} A_{ji}\omega_j.$$

Thus the transpose of the Cartan matrix transforms the fundamental weights into the fundamental roots.

Examples Suppose g has type A_1. Then $A = (2)$ and so $\alpha_1 = 2\omega_1$. Thus $\omega_1 = \frac{1}{2}\alpha_1$.

Now suppose g has type A_2. Then we have

$$A = \begin{pmatrix} 2 & -1 \\ -1 & 2 \end{pmatrix}$$

and so

$$\begin{aligned} \alpha_1 &= 2\omega_1 - \omega_2 \\ \alpha_2 &= -\omega_1 + 2\omega_2. \end{aligned}$$

The fundamental roots and weights are shown in the following figure.

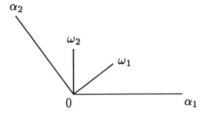

We note that ω_2 is orthogonal to α_1, ω_1 is orthogonal to α_2, and $\omega_1 + \omega_2 = \alpha_1 + \alpha_2$.

3.4 Weyl's character and dimension formulae

Suppose $\lambda \in h^*$ is dominant and integral, so that $L(\lambda)$ is a finite dimensional irreduable g-module. For each $\mu \in h^*$ we define

$$L(\lambda)_\mu = \{v \in L(\lambda); xv = \mu(x)v \text{ for all } x \in h\}$$

The $\mu \in h^*$ for which $L(\lambda)_\mu \neq 0$ are called the weights of $L(\lambda)$. $L(\lambda)_\mu$ is called the μ-weight space of $L(\lambda)$. Its dimension $\dim L(\lambda)_\mu$ is called the multiplicity of the weight μ in $L(\lambda)$. We would like a formula which will enable us to find $\dim L(\lambda)_\mu$ for all μ.

Now all weights μ of $L(\lambda)$ are integral, although not necessarily dominant. Let X be the set of all integral weights. Then

$$X \cong \mathbb{Z} \oplus \ldots \oplus \mathbb{Z}$$

is a free abelian group with basis $\omega_1, \ldots, \omega_l$ of the fundamental weights. Let $\mathbb{Z}X$ be the integral group ring of X. Its elements are finite sums $\sum n_i \lambda_i$ where $n_i \in \mathbb{Z}$ and $\lambda_i \in X$. To give the dimensions of the weight spaces $L(\lambda)_\mu$ is equivalent to giving an element $\sum_{\mu \in X} \dim L(\lambda)_\mu \, \mu$ of $\mathbb{Z}X$.

Now we have a problem in working in the group ring $\mathbb{Z}X$ since we have two types of addition, viz addition in X and addition in $\mathbb{Z}X$. In order to eliminate the confusion arising because of this we define a multiplicative group $e(X)$ isomorphic to the additive group X. $e(X)$ consists of elements $e(\lambda)$ for $\lambda \in X$ where

$$e(\lambda_1)e(\lambda_2) = e(\lambda_1 + \lambda_2).$$

We then work in the group ring $\mathbb{Z}e(X)$. We define the character of $L(\lambda)$ by

$$\operatorname{char} L(\lambda) = \sum_{\mu \in X} \dim L(\lambda)_\mu e(\mu) \in \mathbb{Z}e(X).$$

Now $\mathbb{Z}e(X)$ is an integral domain, so can be embedded in its field of fractions.

H. Weyl determined a formula giving the character of $L(\lambda)$. Let $\rho = \omega_1 + \ldots + \omega_l$. (One can show that ρ is also given by $\rho = 1/2 \sum_{\alpha \in \Phi^+} \alpha$).

Theorem (Weyl's character formula).

$$\operatorname{char} L(\lambda) = \frac{\sum_{w \in W} \det w\, e(w(\lambda + \rho))}{\sum_{w \in W} \det w\, e(w(\rho))}.$$

This is an equality in the field of fractions of $\mathbb{Z}e(X)$. We note that $\det w = \pm 1$ *for each* $w \in W$. *This is because W is generated by reflections s_α and* $\det s_\alpha = -1$ *for each such reflection.*

Example. Suppose that g has type A_1. Then the dominant integral weights are those of form $m\omega_1$, where $m \in \mathbb{Z}$, $m \geq 0$. We consider the character of $L(m\omega_1)$. We have $\rho = \omega_1$. Thus Weyl's character formula gives

$$\operatorname{char} L(m\omega_1) = \frac{\sum_{w \in W} \det w\, e(w(m+1)\omega_1)}{\sum_{w \in W} \det w\, e(w(\omega_1))}.$$

Now the Weyl group W is generated by the single reflection s_{α_1}. Thus $W = \{1, s_{\alpha_1}\}$ has order 2. We have $\det 1 = 1$ and $\det s_{\alpha_1} = -1$. Hence

$$\begin{aligned}\operatorname{char} L(m\omega_1) &= \frac{e((m+1)\omega_1) - e(-(m+1)\omega_1)}{e(\omega_1) - e(-\omega_1)} \\ &= \frac{e(\omega_1)^{m+1} - e(\omega_1)^{-(m+1)}}{e(\omega_1) - e(\omega_1)^{-1}}\end{aligned}$$

Now $\frac{z^{m+1} - z^{-(m+1)}}{z - z^{-1}} = z^m + z^{m-2} + \cdots + z^{-m}$. Hence

$$\begin{aligned}\operatorname{char} L(m\omega_1) &= e(\omega_1)^m + e(\omega_1)^{m-2} + \cdots + e(\omega_1)^{-m} \\ &= e(m\omega_1) + e((m-2)\omega_1) + \cdots + e(-m\omega_1).\end{aligned}$$

3.4 Weyl's character and dimension formulae

Thus $m\omega_1, (m-2)\omega_1, \ldots, -m\omega_1$ are the weights of $L(m\omega_1)$ each occuring with multiplicity 1. In particular we see that

$$\dim L(m\omega_1) = m+1.$$

$$\underset{-m\omega \quad -(m-2)\omega \qquad \qquad (m-2)\omega \quad m\omega}{\vdash\!\!\!\!\dashv\!\!\!\!\dashv\!\!\cdots\cdots\!\!\dashv\!\!\!\!\dashv\!\!\!\!\dashv} \qquad \omega = \omega_1$$

By specialising Weyl's character formula one can obtain a formula for the dimension of $L(\lambda)$.

Theorem. (Weyl's dimension formula)

$$\dim L(\lambda) = \frac{\prod\limits_{\alpha \in \Phi^+} <\alpha, \lambda+\rho>}{\prod\limits_{\alpha \in \Phi^+} <\alpha, \rho>}$$

A slightly different version of this formula is useful for calculating the dimensions in practice. For each positive root $\alpha \in \Phi^+$ we may express α as a linear combinations of the fundamental roots by

$$\alpha = \sum_{i=1}^{l} k_i \alpha_i, \qquad k_i \in \mathbb{Z}, k_i \geq 0.$$

The dominant integral weight λ can be expressed as a combination of the fundamental weights by

$$\lambda = \sum_{i=1}^{l} m_i \omega_i, \qquad m_i \in \mathbb{Z}, m_i \geq 0.$$

Let $\alpha \in \Pi$ be a fundamental root of minimal length. Then we know that for each fundamental root $\alpha_i \in \Pi$ we have

$$<\alpha_i, \alpha_i> = w_i <\alpha, \alpha>$$

where $w_i = 1, 2$ or 3. $w_i = 1$ if α_i is a short root and $w_i = 2$ or 3 if α_i is a long root. $w_i = 3$ only in type G_2. With this notation we have the following corollary of Weyl's dimension formula

$$\dim L(\lambda) = \prod_{\alpha \in \Phi^+} \frac{\sum\limits_{i=1}^{l} k_i w_i (m_i + 1)}{\sum\limits_{i=1}^{l} k_i w_i}.$$

Examples. Let g have type A_1. Let $\lambda = m_1 \omega_1$. Then we have a single positive root α_1 and so

$$\dim L(\lambda) = m_1 + 1.$$

Now let g have type A_2. Let $\lambda = m_1\omega_1 + m_2\omega_2$. This time we have

$$\Phi^+ = \{\alpha_1, \alpha_2, \alpha_1 + \alpha_2\}.$$

We have $w_1 = 1$, $w_2 = 1$ since α_1, α_2 have the same length. Thus

$$\dim L(\lambda) = \frac{(m_1+1)(m_2+1)(m_1+m_2+2)}{2}.$$

The dimensions of some of these irreducible g-modules are shown in the figure

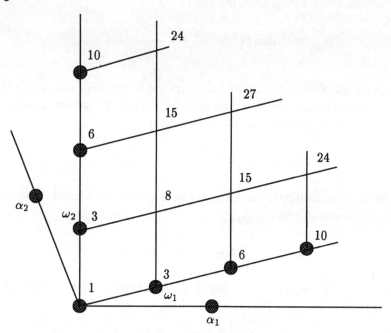

3.5 Fundamental representations

The modules $L(\omega_i)$ $i = 1, \ldots, l$ are called the fundamental irreducible g-modules. If these are known all the others can be obtained as submodules of their tensor products. Thus if $\lambda = m_1\omega_1 + \cdots + m_l\omega_l$ then $L(\lambda)$ is a submodule of

$$\underbrace{L(\omega_1) \otimes \cdots \otimes L(\omega_1)}_{m_1} \otimes \cdots \otimes \underbrace{L(\omega_l) \otimes \cdots \otimes L(\omega_l)}_{m_l}$$

We shall describe the dimensions of the fundamental representations of the simple Lie algebras of classical type A_l, B_l, C_l, D_l. It is convenient to

3.5 Fundamental representations

write these dimensions on the appropriate node of the Dynkin diagram. The dimensions of the fundamental modules are as follows

A_l: $l+1$, $\binom{l+1}{2}$, $\binom{l+1}{3}$, ..., $l+1$

B_l: $2l+1$, $\binom{2l+1}{2}$, $\binom{2l+1}{3}$, ..., $\binom{2l+1}{l-1}$, 2^l

C_l: $2l$, $\binom{2l}{2}-1$, $\binom{2l}{3}-2l$, $\binom{2l}{4}-\binom{2l}{2}$, ..., $\binom{2l}{l-1}-\binom{2l}{l-3}$, $\binom{2l}{l}-\binom{2l}{l-2}$

D_l: $2l$, $\binom{2l}{2}$, $\binom{2l}{3}$, ..., $\binom{2l}{l-2}$, 2^{l-1}, 2^{l-1}

The representation coming from the fundamental module $L(\omega_1)$ is called the basic representation. This representation gives the description of g as an algebra of $n \times n$ matrices where $n = l+1, 2l+1, 2l, 2l$ respectively for types A_l, B_l, C_l, D_l. Most of the other fundamental representations can be obtained by considering exterior powers of the basic representation.

Let V be the module giving the basic representation. Let $T(V)$ be the tensor algebra of V, given by

$$T(V) = \mathbb{C}1 \oplus V \oplus (V \otimes V) \oplus (V \otimes V \otimes V) \oplus \cdots.$$

Let I be the 2-sided ideal of $T(V)$ generated by elements $v \otimes v$ for all $v \in V$. Let $\wedge(V) = T(V)/I$. $\wedge(V)$ is called the exterior algebra of V. We have

$$\wedge(V) = \wedge^0(V) \oplus \wedge^1(V) \oplus \wedge^2(V) \oplus \cdots$$

where $\wedge^i(V)$ is the image of $T^i(V)$. We have

$$\dim \wedge^i(V) = \binom{\dim V}{i}.$$

Let $v_1 \wedge \cdots \wedge v_i \in \wedge^i(V)$ be the image of $v_1 \otimes \cdots \otimes v_i \in T^i(V)$. $\wedge^i(V)$ can be made into a g-module by the rule

$$x(v_1 \wedge \cdots \wedge v_i) = xv_1 \wedge v_2 \wedge \cdots \wedge v_i + \cdots + v_1 \wedge \cdots \wedge v_{i-1} \wedge xv_i$$

for all $x \in g$. We have

$$\wedge^0 V = \mathbb{C}1, \qquad \wedge^1 V \cong V.$$

The $\wedge^i V$ are called the exterior powers of the g-module V.

If g is of type A_l then the exterior powers $\wedge^i V$ for $i = 1, 2, \ldots, l$ give all the fundamental g-modules.

If g is of type B_l the exterior powers $\wedge^i V$ for $i = 1, 2, \ldots, l-1$ give fundamental g-modules. There is one remaining fundamental module not given in this way. $L(\omega_l)$ is called the spin module. It has dimension 2^l and can be constructed from an algebra called the Clifford algebra of V.

If g is of type C_l the exterior powers $\wedge^i V$ of the basic module are not in general irreducible. There exist g-module homomorphisms

$$\theta : \wedge^{i-1} V \longrightarrow \wedge^{i+1} V \quad \text{(expansion)}$$

$$\phi : \wedge^{i+1} V \longrightarrow \wedge^{i-1} V \quad \text{(contraction)}$$

for $1 \leqslant i \leqslant l-1$. θ is injective, ϕ is surjective and we have

$$\wedge^{i+1} V = \ker \phi \oplus \operatorname{im} \theta.$$

The modules $\ker \phi$ for $i = 1, \ldots, l-1$ give the fundamental modules in addition to V.

Now let g have type D_l. Then the exterior powers $\wedge^i V$ for $i = 1, \ldots, l-2$ give fundamental g-modules. This time there are two remaining fundamental modules $L(\omega_{l-1})$, $L(\omega_l)$. They both have dimension 2^{l-1}. They are called the spin modules and are constructed using the Clifford algebra of V.

We shall also give the dimensions of some of the fundamental modules for the simple Lie algebras of exceptional type.

3.5 Fundamental representations

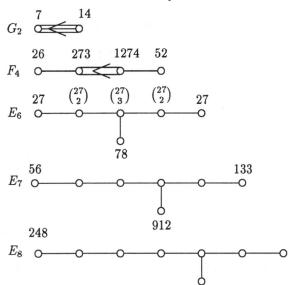

4
Simple groups of Lie type

It is possible to use the theory of simple Lie algebras over $\mathbb{C}$ to construct simple groups of matrices over any field. This was discovered by C. Chevalley. We shall outline Chevalley's theory in the present section.

4.1 A Chevalley basis of g

Let g be a non-trivial simple Lie algebra over $\mathbb{C}$ and h be a Cartan subalgebra of g. Let

$$g = h \oplus \sum_{\alpha \in \Phi} g_\alpha$$

be the Cartan decomposition of g with respect to h. We aim to choose a basis of g adapted to this Cartan decomposition with favourable properties. We first choose a basis of h consisting of the fundamental coroots $h_1, \cdots, h_l$. It is useful to define the coroot $h_\alpha \in h$ corresponding to any root $\alpha \in \Phi$. This is the element of h corresponding to $\frac{2\alpha}{\langle \alpha, \alpha \rangle} \in h^*$ under our isomorphism $h \to h^*$. It can be shown that any coroot h_α is a linear combination of the fundamental coroots $h_1, \cdots, h_l$ with coefficients in $\mathbb{Z}$. We then choose non-zero vectors $e_\alpha \in g_\alpha$. Since $\dim g_\alpha = 1$, e_α is determined up to a scalar. One can show that $[e_\alpha e_{-\alpha}]$ is a non-zero multiple of h_α. Thus we may choose $e_\alpha \in g_\alpha$, $e_{-\alpha} \in g_{-\alpha}$ such that $[e_\alpha e_{-\alpha}] = h_\alpha$.

We next consider the product $[e_\alpha e_\beta]$ where $\alpha + \beta \neq 0$. We have, for $x \in h$,

$$\begin{aligned} [x[e_\alpha e_\beta]] &= [[xe_\alpha]e_\beta] + [e_\alpha[xe_\beta]] \\ &= \alpha(x)[e_\alpha e_\beta] + \beta(x)[e_\alpha e_\beta] \\ &= (\alpha + \beta)(x)[e_\alpha e_\beta] \end{aligned}$$

4.1 A Chevalley basis of g

Thus if $\alpha + \beta \notin \Phi$ we have $[e_\alpha e_\beta] = 0$ and if $\alpha + \beta \in \Phi$ we have $[e_\alpha e_\beta] \in g_{\alpha+\beta}$. We suppose $\alpha + \beta \in \Phi$ and let

$$[e_\alpha e_\beta] = N_{\alpha,\beta} e_{\alpha+\beta}$$

We also have

$$[e_{-\alpha} e_{-\beta}] = N_{-\alpha,-\beta} e_{-\alpha-\beta}$$

One can show that

$$N_{\alpha,\beta} N_{-\alpha,-\beta} = -(p+1)^2$$

where p is the non-negative integer such that $\beta, -\alpha + \beta, -2\alpha + \beta, \cdots, -p\alpha + \beta$ are roots but $-(p+1)\alpha + \beta$ is not a root. Such an integer p certainly exists since the set Φ of roots is finite.

It is in fact possible to choose the vectors $e_\alpha \in g_\alpha$ such

$$[e_\alpha e_\beta] = \pm(p+1) e_{\alpha+\beta}$$

whenever $\alpha + \beta \in \Phi$. The multiplication of basis elements is then given by

$$\begin{aligned}
[h_i h_j] &= 0 \\
[h_i e_\alpha] &= \frac{2 <\alpha_i, \alpha>}{<\alpha_i, \alpha_i>} e_\alpha \\
[e_\alpha e_{-\alpha}] &= h_\alpha, \text{ a } \mathbb{Z}-\text{ combination of } h_1, \cdots, h_l \\
[e_\alpha e_\beta] &= \pm(p+1) e_{\alpha+\beta} \text{ if } \alpha + \beta \in \Phi \\
[e_\alpha e_\beta] &= 0 \text{ if } \alpha + \beta \notin \Phi, \alpha + \beta \neq 0
\end{aligned}$$

Thus the Lie product of any two basis elements is a $\mathbb{Z}$-combination of basis elements.

This kind of basis is called a Chevalley basis. The choice of Chevalley basis is not in general unique.

Now the Chevalley basis described above has an even more favourable property than the fact that the multiplication constants lie in $\mathbb{Z}$. For any $\alpha \in \Phi, \lambda \in \mathbb{C}$, we consider the map

$$\text{ad}(\lambda e_\alpha) : g \to g$$

given by $\text{ad}(\lambda e_\alpha) x = \lambda [e_\alpha x]$. It is not difficult to see that this map is nilpotent, i.e.

$$(\text{ad}(\lambda e_\alpha))^k = 0 \text{ for some } k.$$

We can then form the linear map

$$\exp \operatorname{ad}(\lambda e_\alpha) = 1 + \operatorname{ad}(\lambda e_\alpha) + \frac{\operatorname{ad}(\lambda e_\alpha)^2}{2!} + \cdots$$

One can show that this map is an automorphism of g. (The sum on the right is finite since $\operatorname{ad}(\lambda e_\alpha)$ is nilpotent). The Chevalley basis has the following very favourable property. $\exp \operatorname{ad}(\lambda e_\alpha)$ transforms every element of the Chevalley basis into a linear combination of basis elements with coefficients which are polynomials in λ with coefficients in $\mathbb{Z}$. This is in spite of the denominators appearing in the formula for the exponential! Let $A_\alpha(\lambda)$ be the matrix representing $\exp \operatorname{ad}(\lambda e_\alpha)$ with respect to the Chevalley basis. Then the entries of the matrix $A_\alpha(\lambda)$ are polynomials in λ with coefficients in $\mathbb{Z}$.

Let $G_{\operatorname{ad}}(\mathbb{C})$ be the subgroup of the group of automorphisms of g generated by the elements $\exp \operatorname{ad}(\lambda e_\alpha)$ for all $\alpha \in \Phi$ and all $\lambda \in \mathbb{C}$. This is called the adjoint algebraic group with Lie algebra g. It is a simple group.

We shall now show how one can consider analogous groups over any field.

4.2 Chevalley groups over an arbitrary field

Now let k be any field. Then for each $\alpha \in \Phi$ and each $\mu \in k$ we have a non-singular matrix $A_\alpha(\mu)$ obtained by replacing the indeterminate λ by the element $\mu \in k$. Let $G_{\operatorname{ad}}(k)$ be the group of non-singular matrices over k generated by the matrices $A_\alpha(\mu)$ for all $\alpha \in \Phi$ and all $\mu \in k$. $G_{\operatorname{ad}}(k)$ is called the adjoint Chevalley group of type g over k.

It turns out that the group $G_{\operatorname{ad}}(k)$ is simple, apart from a small finite number of exceptions when k is finite.

Examples

(i) Suppose g is of type A_l. Then $G_{\operatorname{ad}}(k)$ is isomorphic to $PSL_{l+1}(k)$, the projective special linear group of degree $l+1$ over k. We have

$$PSL_{l+1}(k) = SL_{l+1}(k)/Z$$

where Z is the centre of $SL_{l+1}(k)$.

(ii) Suppose g is of type C_l. Then $G_{\operatorname{ad}}(k)$ is isomorphic to $PSp_{2l}(k)$, the projective symplectic group of degree $2l$ over k. We have

$$PSp_{2l}(k) = Sp_{2l}(k)/Z$$

where Z is the centre.

(iii) Suppose g is of type D_l. Then $G_{ad}(k)$ is isomorphic to $P\Omega_{2l}(k, f_D)$. Here

$$P\Omega_{2l}(k, f_D) = \Omega_{2l}(k, f_D)/Z$$

and $\Omega_{2l}(k, f_D)$ is the commutator subgroup of the orthogonal group $O_{2l}(k, f_D)$. This is the group of all non-singular linear transformations of $2l$-dimensional vector space over k leaving invariant the quadratic form f_D with symmetric matrix

$$\begin{pmatrix} 0 & I_l \\ I_l & 0 \end{pmatrix}$$

(iv) Now suppose g is of type B_l. Then $G_{ad}(k)$ is isomorphic to $P\Omega_{2l+1}(k, f_B)$. This time $\Omega_{2l+1}(k, f_B)$ is the commutator subgroup of $O_{2l+1}(k, f_B)$, the orthogonal group of all non-singular linear transformations of $(2l + 1)$-dimensional vector space over k leaving invariant the quadratic form f_B with symmetric matrix

$$\begin{pmatrix} 1 & 0 & \cdots & 0 \\ 0 & & 0 & I_l \\ 0 & & I_l & 0 \end{pmatrix}$$

4.3 Finite Chevalley groups

We now consider the special case when k is a finite field. We recall that the number of elements in any finite field is a prime power, and that for each prime power $q = p^e$ there is just one field F_q, up to isomorphism, with q elements. When $k = F_q$ we shall write $G_{ad}(k) = G_{ad}(q)$.

The number of elements in $G_{ad}(q)$ turns out to be given by an order formula of the following type:

$$|G_{ad}(q)| = \frac{1}{d} q^{|\Phi^+|}(q^{d_1} - 1)(q^{d_2} - 1) \cdots (q^{d_l} - 1).$$

Here d is a small number bounded by an integer independent of q. The numbers $d_1, \cdots, d_l$ are certain positive integers which can be obtained from the root system Φ as follows. For each $\alpha \in \Phi^+$ we can write

$$\alpha = k_1 \alpha_1 + \cdots + k_l \alpha_l \qquad k_i \geqslant 0$$

We define the height of α by

$$\operatorname{ht} \alpha = k_1 + \cdots + k_l.$$

Thus the fundamental roots are the roots of height 1. Suppose there are r_1 roots of height 1, r_2 of height 2, etc. One can show that

$$r_1 \geqslant r_2 \geqslant r_3 \geqslant \cdots$$

and $r_1 + r_2 + \cdots = |\Phi^+|$. Thus we obtain a partition of $|\Phi^+|$. This partition can be represented by a diagram with r_1 squares in row 1, r_2 squares in row 2, etc. For example the diagram of the partition 3221 is

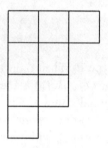

The dual partition is the partition whose parts are the lengths of the columns of this diagram. For example the dual of 3221 is 431. Since $r_1 = l$ the dual of the partition $r_1, r_2, r_3 \cdots$ will have l parts. Let them be $m_1, m_2, \cdots, m_l$. Then the numbers $d_1, \cdots, d_l$ we require are given by $d_i = m_i + 1$.

The group $G_{\mathrm{ad}}(q)$ turns out to be a finite simple group, except in the cases $A_1(2)$, $A_1(3)$, $B_2(2)$, $G_2(2)$. These are called the simple Chevalley groups. Their orders are given in the following table.

$$|A_l(q)| = \frac{q^{l(l+1)/2}}{(l+1, q-1)}(q^2-1)(q^3-1)\cdots(q^{l+1}-1)$$

$$|B_l(q)| = |C_l(q)| = \frac{q^{l^2}}{(2, q-1)}(q^2-1)(q^4-1)\cdots(q^{2l}-1)$$

$$|D_l(q)| = \frac{q^{l(l-1)}}{(4, q^l-1)}(q^2-1)(q^4-1)\cdots(q^{2l}-1)$$

$$|G_2(q)| = q^6(q^2-1)(q^6-1)$$

$$|F_4(q)| = q^{24}(q^2-1)(q^6-1)(q^8-1)(q^{12}-1)$$

$$|E_6(q)| = \frac{q^{36}}{(3, q-1)}(q^2-1)(q^5-1)(q^6-1)(q^8-1)(q^9-1)(q^{12}-1)$$

$$|E_7(q)| = \frac{q^{63}}{(2, q-1)}(q^2-1)(q^6-1)(q^8-1)(q^{10}-1)(q^{12}-1)(q^{14}-1)(q^{18}-1)$$

$$|E_8(q)| = q^{120}(q^2-1)(q^8-1)(q^{12}-1)(q^{14}-1)(q^{18}-1)(q^{20}-1)(q^{24}-1)(q^{30}-1)$$

4.4 Twisted groups

The finite Chevalley groups are not the only finite simple groups obtainable from the Lie theory. There are also twisted groups, obtained independently by R. Steinberg and J. Tits. One can obtain twisted groups whenever the Dynkin diagram (including arrows) has a symmetry.

The possible symmetries are

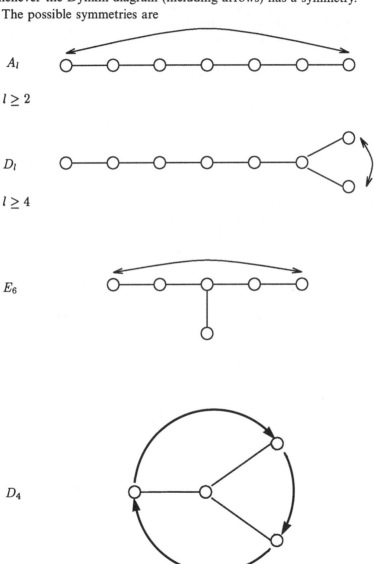

Suppose the Dynkin diagram of g has a symmetry $\alpha \to \bar{\alpha}$. This symmetry has order 2 or 3. We suppose the field k has an automorphism $\lambda \to \bar{\lambda}$ of the same order. We note that a finite field with an automorphism of order 2 must be F_{q^2} with $\bar{\lambda} = \lambda^q$ and a finite field with an automorphism of order 3 must be F_{q^3} with $\bar{\lambda} = \lambda^q$ (or $\bar{\lambda} = \lambda^{q^2}$).

Let U be the subgroup of $G_{ad}(k)$ generated by the matrices $A_\alpha(\lambda)$ for $\alpha \in \Pi$, $\lambda \in k$ and V be the subgroup generated by the $A_\alpha(\lambda)$ for $-\alpha \in \Pi$, $\lambda \in k$. Then $G_{ad}(k)$ is generated by U and V. Now there is an automorphism $\sigma : U \to U$ uniquely determined by

$$\sigma(A_\alpha(\lambda)) = A_{\bar{\alpha}}(\bar{\lambda}) \qquad \alpha \in \Pi, \lambda \in k.$$

Similarly there is an automorphism $\sigma : V \to V$ uniquely determined by

$$\sigma(A_\alpha(\lambda)) = A_{\bar{\alpha}}(\bar{\lambda}) \qquad -\alpha \in \Pi, \lambda \in k.$$

Let $U^\sigma = \{u \in U; \sigma(u) = u\}$
$V^\sigma = \{v \in V; \sigma(v) = v\}$.
Let $G^1(k)$ be the subgroup of $G_{ad}(k)$ generated by U^σ and V^σ. Then $G^1(k)$ turns out to be a simple group, again with few exceptions. In fact there is only one exception, when $g = A_2$ and $k = F_{2^2}$. $G^1(k)$ is called a twisted simple group. We obtain in particular finite simple groups

$$^2A_l(q^2) \ \ l \geqslant 2, \ \ ^2D_l(q^2) \ \ l \geqslant 4, \ \ ^2E_6(q^2), \ \ ^3D_4(q^3).$$

(The top suffix gives the order of the symmetry). Their orders are given by a formula

$$|G^1(k)| = \frac{1}{d^1} q^{|\Phi^+|}(q^{d_1} - \epsilon_1)(q^{d_2} - \epsilon_2) \cdots (q^{d_l} - \epsilon_l)$$

where $\epsilon_1, \cdots, \epsilon_l$ are certain roots of unity. They are the eigenvalues of the symmetry of the Dynkin diagram. In particular we have:

$$|^2A_l(q^2)| = \frac{q^{l(l+1)/2}}{(l+1, q+1)}(q^2 - 1)(q^3 + 1)(q^4 - 1) \cdots (q^{l+1} + (-1)^l)$$

$$|^2D_l(q^2)| = \frac{q^{l(l-1)}}{(4, q^l + 1)}(q^2 - 1)(q^4 - 1)(q^6 - 1) \cdots (q^{2l-2} - 1)(q^l + 1)$$

$$|^2E_6(q^2)| = \frac{q^{36}}{(3, q+1)}(q^2 - 1)(q^3 + 1)(q^6 - 1)(q^8 - 1)(q^9 + 1)(q^{12} - 1)$$

$$|^3D_4(q^3)| = q^{12}(q^2 - 1)(q^6 - 1)(q^4 - \epsilon)(q^4 - \epsilon^2)$$

where $\epsilon = e^{2\pi i/3}$

4.5 Suzuki and Ree groups

There are still further ways of finding finite simple groups from the Lie theory. These arise in those cases where the Dynkin diagram has a symmetry without arrows, but not when the arrows are included. These are the following cases:

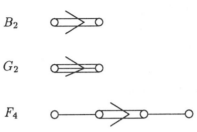

In these cases the symmetry of the diagram does not extend to a symmetry of the root system because the simple roots which correspond under the symmetry do not have the same length. In spite of this, it is still possible to obtain simple groups over certain special fields. In type B_2 this is possible only in characteristic 2. In the finite case one must take a field with 2^{2e+1} elements, i.e an odd power of 2. In type G_2 this is possible only in characteristic 3. In the finite case one must take a field with 3^{2e+1} elements. In type F_4 one has characteristic 2 and a finite field with 2^{2e+1} elements.

In these cases the subgroups U and V have an automorphism uniquely determined by

$$\sigma(A_\alpha(\lambda)) = \begin{cases} A_{\bar\alpha}(\lambda^{p^{e+1}}) & \text{if } \alpha \in \Pi \text{ is short} \\ A_{\bar\alpha}(\lambda^{p^e}) & \text{if } \alpha \in \Pi \text{ is long} \end{cases}$$

$$\sigma(A_\alpha(\lambda)) = \begin{cases} A_{\bar\alpha}(\lambda^{p^{e+1}}) & \text{if } -\alpha \in \Pi \text{ is short} \\ A_{\bar\alpha}(\lambda^{p^e}) & \text{if } -\alpha \in \Pi \text{ is long} \end{cases}$$

(Here $p = 2$ or 3 as appropriate).

Let $U^\sigma = \{u \in U; \sigma(u) = u\}$ and $V^\sigma = \{v \in V; \sigma(v) = v\}$. Let $G^1(k)$ be the subgroup of $G_{\text{ad}}(k)$ generated by U^σ and V^σ. Then $G^1(k)$ is a simple group. It is called a Suzuki group when $g = B_2$ and a Ree group when $g = G_2$ or F_4.

The finite simple groups obtained in this way are

$$\begin{array}{ll} {}^2B_2(2^{2e+1}) & e \geq 1 \\ {}^2G_2(3^{2e+1}) & e \geq 1 \\ {}^2F_4(3^{2e+1}) & e \geq 1. \end{array}$$

It is convenient to write $q^2 = 2^{2e+1}$ or 3^{2e+1} as appropriate. (Thus q is irrational). With this choice of q we have an order formula

$$|G^1(q^2)| = q^{|\Phi^+|}(q^{d_1} - \epsilon_1)(q^{d_2} - \epsilon_2) \cdots (q^{d_l} - \epsilon_l)$$

as before. To be specific we have

$$\begin{aligned} |{}^2B_2(q^2)| &= q^4(q^2 - 1)(q^4 + 1) & q^2 &= 2^{2e+1} \\ |{}^2G_2(q^2)| &= q^6(q^2 - 1)(q^6 + 1) & q^2 &= 3^{2e+1} \\ |{}^2F_4(q^2)| &= q^{24}(q^2 - 1)(q^6 + 1)(q^8 - 1)(q^{12} + 1) & q^2 &= 2^{2e+1} \end{aligned}$$

The Chevalley groups, twisted groups, Suzuki and Ree groups over finite fields are called the finite groups of Lie type.

4.6 Classification of finite simple groups

The classification of finite simple groups was completed in 1981, after many years of intense effort by a number of mathematicians. Every finite simple group is isomorphic to one on the following list:

> A cyclic group of prime order.
> An alternating group of degree $n \geqslant 5$.
> A finite simple group of Lie type.
> A sporadic simple group.

There are 26 sporadic simple groups, of which the largest is the MONSTER. Most of them are subgroups of the MONSTER.

It is interesting to consider to what extent the MONSTER is related to the Lie theory. It is known that the MONSTER is the automorphism group of an infinite dimensional algebra called a vertex operator algebra. Vertex operators appear in the representation theory of the infinite dimensional Lie algebras known as affine Kac-Moody algebras. These are Lie algebras corresponding to the extended Dynkin diagrams on list 2 of §2.5. Thus the MONSTER can be related to the theory of Kac-Moody algebras. Vertex operators are also important in string theory, the branch of theoretical physics which attempts to unify the possible particles and forces. It is intriguing to speculate whether the MONSTER will have a part to play in string theory.

Lie Groups

Graeme Segal

Contents
Lie Groups

Introduction		47
1	**Examples**	49
2	SU_2, SO_3, and $SL_2\mathbb{R}$	53
3	**Homogeneous spaces**	59
4	**Some theorems about matrices**	63
5	**Lie theory**	69
6	**Representation theory**	82
7	**Compact groups and integration**	85
8	**Maximal compact subgroups**	89
9	**The Peter-Weyl theorem**	91
10	**Functions on $\mathbb{R}^n$ and S^{n-1}**	100
11	**Induced representations**	104
12	**The complexification of a compact group**	108
13	**The unitary and symmetric groups**	110
14	**The Borel-Weil theorem**	115
15	**Representations of non-compact groups**	120
16	**Representations of $SL_2\mathbb{R}$**	124
17	**The Heisenberg group**	128

A list of the groups that will be mentioned

The circle group, $\mathbb{T}$, pages 51, 82.

The general and special linear groups:
$GL_n\mathbb{R}$, $GL_n\mathbb{C}$, $SL_n\mathbb{R}$, $SL_n\mathbb{C}$, pages 50, 51, 108.

Othogonal groups:
O_n, SO_n, page 49.
$SO_{1,3}^+$, the Lorentz group, page 54.
$O_n(\mathbb{C})$, page 109.

Unitary groups:
U_n, SU_n, pages 53, 73.
$SU_{1,1}$, page 56.

Symplectic and metaplectic groups:
$Sp_{2n}(\mathbb{R})$, $Mpl_{2n}(\mathbb{R})$, page 131.

The Euclidean group, E_n, page 49.
The Heisenberg group, pages 50, 128, 131.

Introduction

These notes are an expanded version of the seven hours of lectures I gave at Lancaster. I have kept to the original plan and policy, which perhaps need some explanation. Roughly speaking, the contents are what I should like my own graduate students to know about Lie groups, and my general idea was to show how the theory is a natural continuation of basic linear algebra. As root systems and the classification of semisimple Lie algebras were treated in the companion lecture courses I felt I had an excuse for concentrating firmly on the general linear groups. But in any case I believe that is the right way to approach the subject: the taxonomic side of the theory is not to my taste.

I tried to make my lectures useful to people with rather different amounts of mathematical knowledge and sophistication. That means the level is uneven: remarks aimed at the more advanced readers are scattered throughout, and are meant to be ignored by others. I hope the chapters can be read in almost any order: I tried to make them fairly independent. The first four are devoted to a survey of concrete examples of the theory to be developed. This is mainly "undergraduate" material, and so I put it before the formal definition of a Lie group in Chapter 5. But it does not need to be read in advance, and sometimes it uses terminology which is defined only later.

More than half of the book – nearly everything from Chapter 6 on – is concerned with representation theory. I did not at first envisage that this would bulk so large, but in retrospect it does reflect my judgement of what is important. I feel sad that there is nothing about the differential geometry or algebraic topology of Lie groups: I should especially have liked to include the Chern-Weil theory of characteristic classes. I decided, however, that I could not give a worthwhile elementary treatment of these things in the prescribed time. I strongly recommend Milnor's books on Morse Theory and Characteristic Classes to fill the gap.

The text is now at least twice as long as what I actually said, although I have only added "details", and some more proofs. I am not sure the expansion was a good idea: I may well have spoiled the overall perspective by over-egging, while I have certainly not produced a comprehensive treatise.

1
Examples

A good example of a Lie group is the group E_3 of all isometries of euclidean space $\mathbb{R}^3$. Euclidean geometry is the study of those properties of subsets of $\mathbb{R}^3$ which are preserved when the subset is transformed by an element of E_3, so to know what E_3 is is the same thing as to know what is meant by Euclidean geometry. In general, Lie groups are the basic tools of geometry.

Besides being a group a crucial property of E_3 is that it has a topology, i.e. it makes sense to say that one element is "near" another, or to speak of a "continuous path" in E_3. Thus E_3 consists of two connected components, one formed by the elements which preserve orientation and the other by those which reverse it, and there is no continuous path from one of the former to one of the latter.

A simpler example is the subgroup O_3 of E_3 consisting of isometries of $\mathbb{R}^3$ which leave the origin fixed. This can, of course, be identified with the group of 3×3 real orthogonal matrices A. Again it consists of two connected components, the subgroup of matrices A with determinant $+1$, which is called SO_3, and the coset of matrices with determinant -1. The group SO_3 consists of all rotations about axes through the origin in $\mathbb{R}^3$. A rotation is determined by its axis and the angle of rotation, which is taken between 0 and π. Representing a rotation by a vector along its axis whose length is the angle of rotation, and observing that the direction of the axis becomes ambiguous when the angle of rotation is π, one sees that, as a topological space, SO_3 can be made from a solid ball in $\mathbb{R}^3$ of radius π by identifying antipodal points on the boundary sphere. This produces a non-simply-connected space which is not easy to visualize.

Matrix groups

The orthogonal group is an example of a *matrix group*, i.e. a closed subgroup of the group $GL_n\mathbb{R}$ of invertible real $n \times n$ matrices (the composition being, of course, matrix multiplication). All matrix groups are Lie groups. The converse is almost, but not quite, true: all Lie groups are *locally* isomorphic to matrix groups, as will be explained. For the most part the groups we are interested in are matrix groups: the essential reason for preferring the more general concept is that the same group can be realized as a matrix group in many ways, and to make a particular choice often obscures its simplicity and introduces irrelevant features. For example, the additive group $\mathbb{R}$ can be identified with the positive 1×1 matrices (e^x), or with the 2×2 matrices of the form

$$\begin{pmatrix} 1 & x \\ 0 & 1 \end{pmatrix},$$

or with the 2×2 matrices of the form

$$\begin{pmatrix} \cosh x & \sinh x \\ \sinh x & \cosh x \end{pmatrix}.$$

The Euclidean group E_3 is a matrix group because it can be identified with the 4×4 matrices of the form

$$\begin{pmatrix} A & b \\ 0 & 1 \end{pmatrix}$$

with $A \in O_3$ and $b \in \mathbb{R}^3$.

Another reason for considering Lie groups rather than matrix groups is that some groups closely related to matrix groups are not matrix groups. For example, in the group N of 3×3 real upper-triangular matrices with 1's on the diagonal the matrices of the form

$$\begin{pmatrix} 1 & 0 & n \\ 0 & 1 & 0 \\ 0 & 0 & 1 \end{pmatrix}$$

with $n \in \mathbb{Z}$ form a normal subgroup Z. But N/Z is not a matrix group, as we shall prove in on page 83.

This group N/Z can be described in a quite different way. It is called the *Heisenberg group*, and is very important in quantum mechanics. It arises as a group of operators in Hilbert space, i.e., roughly speaking, as

1 Examples

a group of $\infty \times \infty$ matrices. On the Hilbert space $L^2(\mathbb{R})$ let T_a be the operation of translation by a, i.e.

$$(T_a f)(x) = f(x - a).$$

Let M_b be the operation of multiplication by the function $e^{2\pi i b x}$, and let U_c be multiplication by the constant $e^{2\pi i c}$. Then the transformations of $L^2(\mathbb{R})$ of the form $T_a M_b U_c$ form a group which is isomorphic to N/Z.

Low dimensional examples

We can list all the connected Lie groups of dimension ≤ 3.

There are just two connected 1-dimensional groups, $\mathbb{R}$ and $\mathbb{T} = \{z \in \mathbb{C} : |z| = 1\} \cong \mathbb{R}/2\pi\mathbb{Z}$. They are locally isomorphic, in the sense I shall define in a moment.

There is only one connected 2-dimensional group which is not abelian, namely the group of affine transformations $x \longmapsto ax + b$ of the real line, with $a > 0$.

In dimension 3, apart from products of 1- and 2-dimensional groups, the possible groups, up to local isomorphism, are

$$SO_3,\ SL_2\mathbb{R},\ N,\ \text{and}\ E(H),$$

where N is the group of 3×3 upper-triangular matrices already mentioned, $SL_2\mathbb{R}$ is the group of real 2×2 matrices with determinant 1, and $E(H)$ is the subgroup of the group of all affine transformations $x \longmapsto Ax + b$ of $\mathbb{R}^3$ such that A belongs to a prescribed 1-dimensional subgroup H of $GL_2\mathbb{R}$. (Thus when $H = SO_2$ we have the Euclidean group E_2.)

The groups $E(H_1)$ and $E(H_2)$ are isomorphic if and only if H_1 and H_2 are conjugate in $GL_2\mathbb{R}$, and so we have uncountably many non-isomorphic 3-dimensional groups.†

Most of the theory of Lie groups is exemplified by the groups just listed, and to begin with it may be best not to think about any others. The classification of so-called "semisimple" Lie groups by Dynkin diagrams is enormously important in many areas of mathematics, but it is not very relevant to the kind of questions I shall be concerned with. For

---

† The elements of H can be written e^{tB} for a fixed 2×2 matrix B and $t \in \mathbb{R}$. The ratio of the eigenvalue of B does not change when H is conjugated.

our purposes, it tells us that any semisimple group is an interlocking system of copies of $SL_2\mathbb{R}$ and SO_3. The way the copies interlock can be described purely combinatorially by the techniques of root systems, which are treated in the accompanying lectures on Lie algebras.

Local isomorphism

Two groups G_1 and G_2 are *locally isomorphic* if there is a homeomorphism $f : U_1 \longrightarrow U_2$ between neighbourhoods of the identity elements in the respective groups which preserves the composition law in the sense that $f(xy) = f(x)f(y)$ whenever xy belongs to U_1.

The most obvious locally isomorphic groups are $\mathbb{R}$ and $\mathbb{T}$: we can take $f(x) = e^{ix}$ for $|x| < \pi$. A much more interesting example is the local isomorphism between SU_2 and SO_3 which I shall describe in the next section.

2
SU_2, SO_3, and $SL_2\mathbb{R}$

The group of 2×2 unitary matrices with determinant 1 is denoted by SU_2. There is a homomorphism $SU_2 \to SO_3$ which is 2-to-1 and onto. It is of enormous importance in particle physics, because, while SO_3 can be regarded as the set of possible positions of a rigid body whose centre of mass is fixed at the origin, the set of states of an electron which is at rest at the origin is SU_2. The electron has two states for each way of orienting it in space, and one can change it from one to the other by rotating it through 2π about any axis. (See page 76.)

The elements of the group SU_2 can be written

$$\begin{pmatrix} a & b \\ -\bar{b} & \bar{a} \end{pmatrix} \tag{2.1}$$

where a, b are complex numbers such that $|a|^2 + |b|^2 = 1$. This is the same as the group of *unit quaternions*, i.e. quaternions $q = t + xi + yj + zk$ with t, x, y, z real and $t^2 + x^2 + y^2 + z^2 = 1$. For quaternions can be identified with 2×2 complex matrices of the form (2.1) by

$$i \leftrightarrow \begin{pmatrix} i & 0 \\ 0 & -i \end{pmatrix}, \quad j \leftrightarrow \begin{pmatrix} 0 & 1 \\ -1 & 0 \end{pmatrix}, \quad k \leftrightarrow \begin{pmatrix} 0 & i \\ i & 0 \end{pmatrix}.$$

Thus SU_2 is topologically a 3-dimensional sphere, easier to visualize than SO_3.

Quaternions were invented to describe rotations. Thinking of a quaternion as a real part plus a vector part, i.e. $q = t + v$ with $t \in \mathbb{R}$ and $v \in \mathbb{R}^3$, quaternion multiplication is defined by

$$(t_1 + v_1)(t_2 + v_2) = (t_1 t_2 - \langle v_1, v_2 \rangle) + (t_1 v_2 + t_2 v_1 + v_1 \times v_2),$$

where $v_1 \times v_2$ denotes the usual vector product in three dimensions.

In terms of matrices,

$$\mathbb{R}^3 \leftrightarrow \{\text{skew hermitian matrices with trace 0}\}.$$

For any non-zero quaternion g

$$v \in \mathbb{R}^3 \Rightarrow gvg^{-1} \in \mathbb{R}^3,$$

so each $g \in SU_2$ defines a linear transformation T_g of $\mathbb{R}^3$ by $v \mapsto gvg^{-1}$. If $u \in \mathbb{R}^3$ is a unit vector then

$$g = \cos\theta + u\sin\theta$$

is a unit quaternion, and $T_g : \mathbb{R}^3 \to \mathbb{R}^3$ is rotation about the axis u through the angle 2θ. So $g \mapsto T_g$ is a surjective homomorphism

$$T : SU_2 \longrightarrow SO_3.$$

It is easy to check that the kernel of T consists of the two elements ± 1: a rotation is represented not by one quaternion but by a pair $\pm g$.

Thus SU_2 is a two-sheeted covering of SO_3, and is non-trivial in the sense that there is no continuous choice of a single quaternion representative for each rotation. For if $\{R_{u,\theta}\}_{0 \leqslant \theta \leqslant 2\pi}$ is the closed path in SO_3 consisting of rotations through θ about u, and we choose $1 \in SU_2$ to represent $R_{u,0}$, then we must choose $g_\theta = \cos\frac{\theta}{2} + u\sin\frac{\theta}{2}$ to represent $R_{u,\theta}$. But then $R_{u,2\pi}$ is represented by $\cos\pi + u\sin\pi = -1 \neq 1$, and the path $\{g_\theta\}$ in SU_2 does not close.

There are a number of closely related double-covering homomorphisms worth mentioning at this point.

(i) Thinking of the quaternions as $\mathbb{R}^4$, an arbitary element of SO_4 can be written $v \mapsto g_1 v g_2^{-1}$, where g_1 and g_2 are unit quaternions. This gives a double covering

$$SU_2 \times SU_2 \longrightarrow SO_4.$$

(ii) The homomorphism $T : SU_2 \to SO_3$ extends to a double covering

$$T : SL_2\mathbb{C} \to SO_{1,3}^+,$$

where $SO_{1,3}^+$ is the *Lorentz group* (i.e. the elements of $SL_4(\mathbb{R})$ which preserve the quadratic form $t^2 - x^2 - y^2 - z^2$ and also preserve the direction of time, i.e. do not interchange the two sheets of the hyperboloid

$t^2 - x^2 - y^2 - z^2 = 1$). To see this, we identify $\mathbb{R}^4$ with the 2×2 hermitian matrices by

$$(t, x, y, z) \rightarrow \begin{pmatrix} t+x & y-iz \\ y+iz & t-x \end{pmatrix},$$

and let $g \in SL_2\mathbb{C}$ act on them by $A \mapsto gAg^{-1}$. The quadratic form is preserved because

$$\det \begin{pmatrix} t+x & y-iz \\ y+iz & t-x \end{pmatrix} = t^2 - x^2 - y^2 - z^2.$$

(iii) Restricting $T : SL_2\mathbb{C} \to SO_{1,3}^+$ to real matrices, we get a double covering

$$T : SL_2\mathbb{R} \to SO_{1,2}^+.$$

(iv) Finally, the covering $SU_2 \times SU_2 \to SO_4$ defines a double covering

$$\Pi : SO_4 \to SO_3 \times SO_3,$$

for the two elements of $SU_2 \times SU_2$ above $g \in SO_4$ have the same image in $SO_3 \times SO_3$. The homomorphism Π describes the action of SO_4 on $\Lambda^2 \mathbb{R}^4$, which splits into two 3-dimensional pieces $\Lambda_+^2 \oplus \Lambda_-^2$, the *self-dual* and *anti-self-dual* parts, under the action of SO_4.

The relation between SU_2 and SO_3, and between $SL_2\mathbb{C}$ and $SO_{1,3}^+$ is so important that it is worth giving an alternative description of it.

Think of the unit sphere S^2 in $\mathbb{R}^3$ as the Riemann sphere $\Sigma = \mathbb{C} \cup \{\infty\}$ by stereographic projection, i.e.

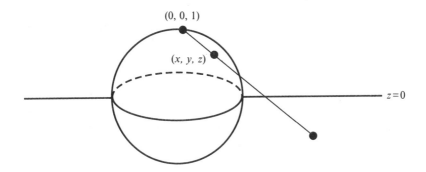

$$(x, y, z) \in S^2 \longleftrightarrow \xi + i\eta = \frac{x+iy}{1-z} \in \Sigma.$$

To
$$g = \begin{pmatrix} a & b \\ -\bar{b} & \bar{a} \end{pmatrix}$$
in SU_2 we associate the Möbius transformation
$$z \mapsto \frac{az + b}{-\bar{b}z + \bar{a}}.$$

This is a bijection $\Sigma \to \Sigma$, which, when regarded as a map $S^2 \to S^2$, is precisely the rotation Tg.

Any holomorphic bijection $\Sigma \to \Sigma$ is a Möbius transformation $z \mapsto (az + b)/(cz + d)$, and one may as well assume that $ad - bc = 1$. Changing the sign of a, b, c, d is immaterial, so the group of Möbius transformations is the quotient group of $SL_2\mathbb{C}$ by its centre, which consists of the matrices ± 1. This group is denoted by $PSL_2\mathbb{C}$.

The sphere S^2 can be regarded as the "celestial sphere", i.e. the set of light rays through the origin in Minkowski space. Thus the Lorentz group acts as a group of transformations of S^2. The surjection
$$SL_2\mathbb{C} \to SO_{1,3}^+$$
tells us that the Möbius transformations of Σ are precisely the Lorentz transformations of the celestial sphere S^2. One way of looking at this, emphasised by Roger Penrose, is to say that the celestial sphere is naturally a 1-dimensional *complex* manifold, and the Lorentz group is the group of all holomorphic bijections $S^2 \to S^2$.

A picture of $SL_2\mathbb{R}$

To visualize $SL_2\mathbb{R}$ it helps to notice that it is isomorphic to the group of complex matrices of the form
$$\begin{pmatrix} a & b \\ \bar{b} & \bar{a} \end{pmatrix}$$
such that $|a|^2 - |b|^2 = 1$. This group is called $SU_{1,1}$. In fact $SL_2\mathbb{R}$ and $SU_{1,1}$ are conjugate subgroups in $GL_2\mathbb{C}$, for $g(SL_2\mathbb{R})g^{-1} = SU_{1,1}$, where
$$g = \begin{pmatrix} 1 & -i \\ 1 & i \end{pmatrix}.$$

The group $SL_2\mathbb{R}$ corresponds to the Möbius transformations $z \mapsto (az + b)/(cz + d)$ of the Riemann sphere $\mathbb{C} \cup \{\infty\}$ which preserve the

2 SU_2, SO_3, and $SL_2\mathbb{R}$

upper half-plane $H = \{z \in \mathbb{C} : \text{Im}(z) > 0\}$, while $SU_{1,1}$ corresponds to those which preserve the disc $D = \{z \in \mathbb{C} : |z| < 1\}$. The transformation $z \mapsto (z-i)/(z+i)$ defined by g takes H to D.

The group $SU_{1,1}$ is homeomorphic to an open solid torus $S^1 \times D$ by

$$\begin{pmatrix} a & b \\ \bar{b} & \bar{a} \end{pmatrix} \longleftrightarrow (a/|a|, b/a) \in S^1 \times D.$$

Whether regarded as $SL_2\mathbb{R}$ or as $SU_{1,1}$, the group has three kinds of elements apart from the two elements ± 1 which form the centre:

(i) those with $|\text{trace}| > 2$, called *hyperbolic*, which in $SL_2\mathbb{C}$ are conjugate to $\begin{pmatrix} \lambda & 0 \\ 0 & \lambda^{-1} \end{pmatrix}$ for some $\lambda \in \mathbb{R}$;

(ii) those with $|\text{trace}| < 2$, called *elliptic*, which are conjugate to $\begin{pmatrix} e^{i\alpha} & 0 \\ 0 & e^{-i\alpha} \end{pmatrix}$ for some $\alpha \in \mathbb{R}$;

(iii) the interface, with $|\text{trace}| = 2$, called *parabolic*, conjugate to $\pm \begin{pmatrix} 1 & 1 \\ 0 & 1 \end{pmatrix}$.

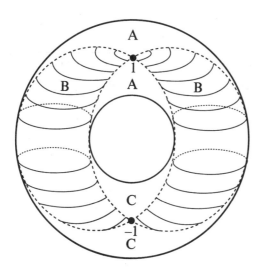

The elliptic elements, which form the sausage-like region B, are the union of all subgoups of $SL_2\mathbb{R}$ which are isomorphic to the circle-group $\mathbb{T}$. The closure of the region A consists of the elements with trace ≥ 2. It is the union of all subgroups isomorphic to $\mathbb{R}$. The region C is the elements with trace < -2. These do not belong to any 1-parameter subgroup. (See page 74.)

The two kinds of 1-parameter subgroups in $SL_2\mathbb{R}$ are related to the positive and negative values of the *Killing form* (see page 15). A neighbourhood of the identity element in $SL_2\mathbb{R}$ can be identified with a neighbourhood of 0 in the vector space $\mathfrak{g}$ of 2×2 matrices of the form

$$\begin{pmatrix} a & b+c \\ b-c & -a \end{pmatrix}$$

by the exponential map (see page 73). In this notation, the Killing form on $\mathfrak{g}$ is $8(a^2 + b^2 - c^2)$, and the regions A and B correspond to the parts of $\mathfrak{g}$ where the Killing form is positive and negative. We see here the simplest case of a general fact. A *semisimple* group is one for which the Killing form is non-degenerate. In a matrix group of this type the Lie algebra of a maximal compact subgroup is a maximal subspace of $\mathfrak{g}$ on which the Killing form is *negative-definite*.

The simply connected covering group of $SL_2\mathbb{R}$ is the infinite open cylinder got by unwrapping the torus $S^1 \times D$.

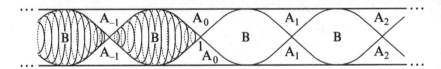

It is homeomorphic to $\mathbb{R}^3$, and is *not* a matrix group (see page 130). The elements in the regions A_0 and B belong to subgroups isomorphic to $\mathbb{R}$, while those in A_k for $k \neq 0$ do not belong to any 1-parameter subgroup.

3
Homogeneous spaces

Lie groups arise as transformation groups. Spaces on which a Lie group acts transitively are known as *homogeneous spaces*. For example,

(i) the sphere $S^{n-1} = \{x \in \mathbb{R}^n : \|x\| = 1\}$ is homogeneous under O_n;

(ii) the upper half-plane $H = \{z \in \mathbb{C} : \text{Im}(z) > 0\}$ is homogeneous under $SL_2\mathbb{R}$, acting by $z \mapsto (az+b)/(cz+d)$;

(iii) the space $\mathscr{P}$ of positive-definite real symmetric $n \times n$ matrices is homogeneous under $GL_n\mathbb{R}$, acting by $(A, P) \mapsto APA^t$;

(iv) the *Grassmannian* $\text{Gr}_k(\mathbb{R}^n)$, defined as the set of all k-dimensional vector subspaces of $\mathbb{R}^n$, is homogeneous under the action of $GL_n\mathbb{R}$, but also homogeneous under the subgroup O_n, because any point of $\text{Gr}_k(\mathbb{R}^n)$ has an *orthonormal* basis;

(v) the space $\mathscr{L}$ of *lattices* in $\mathbb{R}^2$ — a lattice is a subgroup isomorphic to $\mathbb{Z} \oplus \mathbb{Z}$ generated by a basis $\{v_1, v_2\}$ of $\mathbb{R}^2$ — is homogeneous under $GL_2\mathbb{R}$; and the subspace $\mathscr{L}_1$ of *unimodular* lattices (those where the basic parallelogram $\{v_1, v_2\}$ has unit area) is homogeneous under $SL_2\mathbb{R}$.

When a group G acts transitively on a set X we can identify X with the set G/H of left-cosets of the isotropy group H of a point $x_0 \in X$. (To be precise, $H = \{g \in G : gx_0 = x_0\}$, and the map $G/H \xrightarrow{\cong} X$ is $gH \mapsto gx_0$.)

In the five examples above we get

(i) $S^{n-1} \cong O_n/O_{n-1}$ by taking $x_0 = e_n$, the nth basis vector of $\mathbb{R}^n$;

(ii) $H \cong SL_2\mathbb{R}/SO_2$ by taking $x_0 = i$;

(iii) $\mathscr{P} \cong GL_n\mathbb{R}/O_n$, by taking $x_0 = 1$;

(iv) $\text{Gr}_k(\mathbb{R}^n) \cong GL_n\mathbb{R}/GL_{k,n-k} \cong O_n/(O_k \times O_{n-k})$, where $GL_{k,n-k}$ is the group of echelon matrices $\binom{**}{0*}$;

(v) $\mathscr{L} \cong GL_2\mathbb{R}/GL_2\mathbb{Z}$; $\mathscr{L}_1 \cong SL_2\mathbb{R}/SL_2\mathbb{Z}$.

So far in this section we have ignored topology. But in fact each of the spaces $S^{n-1}, H, \mathscr{P}$, and $\text{Gr}_k(\mathbb{R}^n)$ has a natural topology (for $\text{Gr}_k(\mathbb{R}^n)$ see page 71), and the isomorphisms just listed are homeomorphisms between the natural topology and the topology which the set of cosets G/H acquires as a quotient space of G. The proof is in each case an easy exercise.

The spaces of lattices $\mathscr{L}$ and $\mathscr{L}_1$ have several remarkable and unobvious descriptions. It turns out that $\mathscr{L}_1$ is homeomorphic to the complement of a trefoil knot in $\mathbb{R}^3$

(see [Milnor][2] page 84), while $\mathscr{L}$ is homeomorphic to the space of unordered triples of distinct points in the plane $\mathbb{R}^2$, with centre of mass at the origin.

Symmetric spaces

If X is a Riemannian manifold the group of isometries of X is always a Lie group. For, because an isometry $f : X \to X$ preserves geodesics, it is completely determined by $f(x_0)$ and $Df(x_0)$, where x_0 is a base-point in X.

A connected Riemannian manifold X is called a *symmetric space* if for each $x \in X$ there is an isometry $f_x : X \to X$ which reverses geodesics through x, i.e. is such that $Df_x(x) = -1$. A symmetric space is always homogeneous, for any two points x and y can always be joined by a geodesic γ, and $f_z(x) = y$ if z is the mid-point of γ. In fact $X \cong G/H$, where H is the subgroup of G left fixed by an automorphism α of G such that $\alpha^2 = 1$.

Symmetric spaces can be completely classified, and are of great importance in geometry. (See the book by Helgason.) The spaces in examples (i) to (iv) above are symmetric.

For future reference (see page 119) we shall mention another important symmetric space.

Complex structures on $\mathbb{R}^{2n}$

Let $\mathscr{J}_n$ be the space of complex structures on $\mathbb{R}^{2n}$ which are compatible with the inner product, i.e.

$$\mathscr{J}_n = \{J \in O_{2n} : J^2 = -1\}.$$

For any $J \in \mathscr{J}_n$ it is clear that we can find an orthonormal basis $\{v_i\}$ of $\mathbb{R}^{2n}$ such that

$$Jv_{2k-1} = v_{2k} \text{ and } Jv_{2k} = -v_{2k-1}.$$

So any two Js are conjugate in O_{2n}, and $\mathscr{J}_n$ is the homogeneous space O_{2n}/U_n.

The space $\mathscr{J}_n$ can also be identified with the *isotropic Grassmannian* of $\mathbb{C}^{2n}$, i.e. the set of n-dimensional complex subspaces W of $\mathbb{C}^{2n}$ such that $W = W^\perp$ with respect to the $\mathbb{C}$-bilinear extension of the inner-product of $\mathbb{R}^{2n}$. For a complex structure J is the same thing as a splitting $\mathbb{C}^{2n} = W \oplus \bar{W}$ into isotropic subspaces, where W and $\bar{W}$ are the $(\pm i)$-eigenspaces of J.

This description of $\mathscr{J}_n$ shows that it is a *complex* manifold. In fact it is a homogeous space of $O_{2n}(\mathbb{C})$. To see this we choose a basis of $\mathbb{C}^{2n}$ of the form $\{u_1, \ldots, u_n; \bar{u}_1, \ldots \bar{u}_n\}$, where

$$\begin{aligned} \langle u_i, u_j \rangle &= \langle \bar{u}_i, \bar{u}_j \rangle = 0 \text{ and} \\ \langle u_i, \bar{u}_j \rangle &= \delta_{ij}. \end{aligned}$$

(For example, take $u_k = 2^{-\frac{1}{2}}(e_{2k-1} + ie_{2k})$, where $\{e_i\}$ is the usual basis.) Then $O_{2n}(\mathbb{C})$ consists of the complex matrices g such that $g^t A g = A$, where A is the block matrix

$$\begin{pmatrix} 0 & 1 \\ 1 & 0 \end{pmatrix},$$

and O_{2n} is the subgroup of matrices of the form

$$\begin{pmatrix} a & \bar{c} \\ c & \bar{a} \end{pmatrix}.$$

Let P be the subgroup of $O_{2n}(\mathbb{C})$ consisting of matrices of the form

$$\begin{pmatrix} a & b \\ 0 & \check{a} \end{pmatrix},$$

where $\check{a}$ denotes $(a^t)^{-1}$ and $a^{-1}b$ is skew. It is easy to check that $P \cap O_{2n} = U_n$, and $O_{2n}/U_n \cong O_{2n}(\mathbb{C})/P$.

Finally, it should be mentioned that alongside $\mathscr{J}_n$ there is another symmetric space $\mathscr{H}_n = Sp_{2n}(\mathbb{R})/U_n$ formed by the complex structures on $\mathbb{R}^{2n}$ which preserve a *skew* rather than a symmetric bilinear form. The space $\mathscr{H}_n$ is called the *Siegel generalized upper half-plane*.

4
Some theorems about matrices

In this section I shall recall four well-known theorems about matrices which have important generalizations as theorems about Lie groups. The first three describe canonical ways of factorizing a general invertible matrix.

A The polar decomposition

Theorem 4.1 *Any invertible real $n \times n$ matrix g has a unique factorization $g = pu$, where p is a positive-definite symmetric matrix, and u is orthogonal.*

Proof. One defines $p = (gg^t)^{\frac{1}{2}}$ and $u = p^{-1}g$, observing that the positive-definite symmetric matrix gg^t has a unique positive-definite square-root.

The positive-definite symmetric matrices form a convex open subset $\mathscr{P}$ in the vector space $\mathfrak{p}$ of symmetric matrices, so $\mathscr{P}$ is homeomorphic to $\mathbb{R}^{\frac{1}{2}n(n+1)}$, and the theorem implies that $GL_n\mathbb{R}$ is homeomorphic to the product space $\mathscr{P} \times O_n$. The elements of $\mathscr{P}$ do not form a group, but they are precisely the exponentials of the elements of $\mathfrak{p}$.

Theorem (4.1) generalizes from $GL_n\mathbb{R}$ to any Lie group G with finitely many connected components. The subgroup O_n is characterized as a maximal compact subgroup K of G, and $\mathscr{P} = \exp \mathfrak{p}$, where $\mathfrak{p}$ is the orthogonal complement, with respect to the Killing form (see page 15), of the Lie algebra of K in that of G. We shall return to this in Chapter 8.

B The Gram-Schmidt process

There is an algorithm for replacing an arbitrary basis $\{v_1, \ldots v_n\}$ of $\mathbb{C}^n$ by an orthonormal basis $\{u_1, \ldots, u_n\}$. For each k in turn one subtracts a linear combination of $u_1,\ldots,u_{k-1}$ from v_k to obtain a vector $\tilde{v}_k$ which is orthogonal to $u_1,\ldots,u_{k-1}$. Then the vectors $u_k = \tilde{v}_k/\|\tilde{v}_k\|$ form an orthonormal basis of $\mathbb{C}^n$, and we have

$$\begin{aligned} v_1 &= \lambda_{11}u_1 \\ v_2 &= \lambda_{12}u_1 + \lambda_{22}u_2 \\ v_3 &= \lambda_{13}u_1 + \lambda_{23}u_2 + \lambda_{33}u_3, \end{aligned}$$

and so on.

If the vectors $\{v_i\}$ are the columns of an element g of $GL_n\mathbb{C}$ then the $\{u_i\}$ are the columns of a unitary matrix u, and $g = ub$, where b is the upper-triangular matrix with entries (λ_{ij}). Thus we have proved

Theorem 4.2 *Any $g \in GL_n\mathbb{C}$ can be factorized uniquely $g = ub$, where $u \in U_n$ and b is an upper-triangular matrix with positive real diagonal entries.*

If B is the group of all $n \times n$ complex upper-triangular matrices then $U_n \cap B = T$, where $T \cong \mathbb{T}^n$ is the subgroup of diagonal matrices in U_n. So (4.2) implies

Theorem 4.3 *The natural map of homogeneous spaces*

$$U_n/T \longrightarrow GL_n\mathbb{C}/B$$

is a homeomorphism.

The homogeneous space $U_n/T \cong GL_n\mathbb{C}/B$ is very important in the representation theory of U_n and $GL_n\mathbb{C}$. It is the space of *flags* in $\mathbb{C}^n$: a flag is a sequence of subspaces

$$E_1 \subset E_2 \subset \ldots \subset E_n = \mathbb{C}^n$$

with $\dim(E_k) = k$. See page 174.

There is a corresponding theorem for any linear algebraic group: we shall meet it in Chapter 14.

C Reduced echelon form: the Bruhat decomposition

Once again, let $\{v_1,\ldots,v_n\}$ be the basis of $\mathbb{C}^n$ formed by the columns of an invertible matrix g, and let us perform the same kinds of column operations as were used in the Gram-Schmidt process, i.e. multiplying a column by a scalar, and subtracting from it a multiple of any column to its left. Thus once again we are really trying to find a basis for a *flag*. But this time, instead of trying to make the basis orthonormal, we construct the unique basis $w = \{w_1, \ldots, w_n\}$ which is in *reduced echelon form*, i.e. like

$$w = \begin{pmatrix} * & 1 & 0 & 0 \\ * & 0 & * & 1 \\ 1 & 0 & 0 & 0 \\ 0 & 0 & 1 & 0 \end{pmatrix},$$

in which
 (i) each column w_i ends in a 1, say in the π_i^{th} row, and
 (ii) the entries to the right of each 1 vanish, i.e. $w_{\pi_i j} = 0$ if $j > i$.
The sequence

$$\pi = (\pi_1, \ldots, \pi_n)$$

is necessarily a permutation of $(1, 2, \ldots, n)$, and the matrix w is got by permuting the columns of an upper-triangular matrix

$$n = \begin{pmatrix} 1 & 0 & * & 0 \\ 0 & 1 & * & * \\ 0 & 0 & 1 & 0 \\ 0 & 0 & 0 & 1 \end{pmatrix}$$

by π. In fact $w = n\pi$, where the permutation π is identified with the corresponding $n \times n$ permutation-matrix. We have proved

Theorem 4.4 *Any element* $g \in GL_n\mathbb{C}$ *can be factorized* $g = n\pi b$, *where n belongs to the subgroup N of upper-triangular matrices with 1's on the diagonal, π is a permutation matrix, and b belongs to the subgroup B of upper-triangular matrices.*

Equivalently, the permutation matrices π form a set of representatives for the orbits of the action of N on the homogeneous space $GL_n\mathbb{C}/B$.

As expressed in (4.4) the factorization $g = n\pi b$ is not unique, though the permutation π is uniquely determined by g. In fact the conditions (i) and (ii) characterizing w above can be reformulated as
 (i)' $w\pi^{-1} \in N$
 (ii)' $pi^{-1}w \in \tilde{N}$
where $\tilde{N}$ is the group of lower-triangular matrices with 1's on the diagonal. So we have

Theorem 4.5 *The decomposition $g = n\pi b$ is unique if n is chosen in $N_\pi = N \cap \pi \tilde{N} \pi^{-1}$.*

Equivalently, the orbit of $\pi B \in GL_n\mathbb{C}/B$ under N is isomorphic to N_π.

The number of variable entries in the matrix w above is l_π, the *length* of π, which is defined as the number of pairs $i < j$ such that $\pi_j > \pi_j$, i.e. as the number of *crossings* when π is depicted in the form

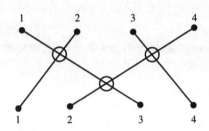

$$\pi = \{3, 1, 4, 2\} \Longrightarrow l_\pi = 3$$

This means that N_π is homeomorphic to the vector space $\mathbb{C}^{l_\pi}$, giving us

Corollary 4.6 *The orbits of N on the flag-manifold $GL_n\mathbb{C}/B$ decompose it into $n!$ cells C_π, with $C_\pi \cong N_\pi \cong \mathbb{C}^{l_\pi}$.*

For future use let us notice that for *almost* all $g \in GL_n\mathbb{C}$ we shall have $\pi = \{n, n-1, n-2, \ldots, 1\}$ and $l_\pi = \frac{1}{2}n(n-1)$. In this case $n\pi \in \pi\tilde{N}$, and, replacing g by πg, we have

Corollary 4.7
Almost all $g \in GL_n\mathbb{C}$ have a unique factorization $g = \tilde{n}b$ with $\tilde{n} \in \tilde{N}$ and $b \in B$.

Everything in this section can be generalized from $GL_n\mathbb{C}$ to any reductive Lie group over any field, even a finite field. In the general case the decomposition is called the *Bruhat decomposition*. The role of the permutation group is taken over by the *Weyl group* of G. (see page 16)

It is instructive to consider the case $G = GL_n\mathbb{F}_q$, where $\mathbb{F}_q$ is a finite field with q elements. Then G has order

$$(q^n - 1)(q^n - q)\ldots(q^n - q^{n-1}) \ ,$$

the upper-triangular subgroup B has order $(q-1)^n q^{\frac{1}{2}n(n-1)}$, and N_π has order q^{l_π}. The Bruhat decomposition gives us the identity

$$|G/B| = \sum_{\pi \in S_n} |N_\pi|,$$

i.e.

$$\prod_{k=1}^{n} \frac{q^k - 1}{q - 1} = \sum_{\pi \in S_n} q^{l_\pi}.$$

D Diagonalization and maximal tori

In the unitary group U_n each element g is conjugate to a diagonal matrix. The diagonal matrices in U_n form a *torus* $T \cong \mathbb{T}^n$, and T is a maximal abelian subgroup of U_n. In fact any abelian subgroup of U_n is conjugate to a subgroup of T. These facts are proved by elementary linear algebra.

No such simple statements can be made about $GL_n\mathbb{C}$. But for any *compact* Lie group G the picture is much as for U_n. We can always choose a maximal torus T in G, i.e. a maximal subgroup of the form $\mathbb{T}^n$.

Theorem 4.8 *If G is a compact connected Lie group then*

(i) *any element of G is conjugate to an element of T, and more generally,*

(ii) *any connected abelian subgroup of G is conjugate to a subgroup of T.*

In particular, any two maximal tori are conjugate.

The word "connected" cannot be omitted in (ii): not every maximal abelian subgroup is a torus.

Example. SO_2 is a maximal torus of SO_3, and the statement (i) amounts to the fact that any element of SO_3 is a rotation about some axis. The diagonal matrices in SO_3 form a maximal abelian subgroup A with four elements, and A is clearly not isomorphic to a subgroup of the circle-group SO_2.

There are various ways to prove Theorem (4.8). The crucial part is (i), for (ii) follows easily from (i) because any compact connected abelian group A contains an element g whose powers are dense in A (see [Adams] page 79), and then

$$x^{-1}gx \in T \Longrightarrow x^{-1}Ax \subset T.$$

There is a very attractive proof of (i) by means of algebraic topology. Again I shall refer to [Adams] (page 90) for more details, but I shall describe the idea.

We first reformulate the assertion as a fixed-point theorem: to find $x \in G$ such that $x^{-1}gx \in T$ is the same as to find a fixed-point of the map $f_g : G/T \to G/T$ defined by $f_g(xT) = gxT$. The map f_g depends continuously on $g \in G$, and G is connected, so f_g is homotopic to the identity-map f_1 of G/T. We now use a well-known theorem of topology.

Theorem 4.9 *If X is a compact space with non-zero Euler number then any map $f : X \to X$ which is homotopic to the identity has a fixed point.*

The *Euler number* is an integer-valued topological invariant $\chi(X)$ defined for compact spaces X which possess a decomposition into cells. It is characterized by three properties.

   (i) $\chi(X) = 0$ if X is empty.
  (ii) $\chi(X) = 1$ if X is contractible.
 (iii) $\chi(X_1 \cup X_2) = \chi(X_1) + \chi(X_2) - \chi(X_1 \cap X_2)$.

By decomposing the n-dimensional sphere S^n as the union of two hemispheres we find, by induction on n, that

$$\chi(S^n) = \begin{cases} 2 & \text{if } n \text{ is even} \\ 0 & \text{if } n \text{ is odd.} \end{cases}$$

(The fact that any map $S^2 \to S^2$ which is homotopic to the identity has a fixed point is the "hairy ball" theorem, probably the most famous result of elementary topology.) A similar argument shows that if X has a cellular decomposition with c_k cells of dimension k then

$$\chi(X) = \Sigma(-1)^k c_k.$$

To prove Theorem (4.8) we need to know that $\chi(G/T)$ is non-zero. The Bruhat decomposition of G/T into even-dimensional cells tells us that $\chi(G/T)$ is the order of the Weyl group of G, but there are easier proofs.

5
Lie theory

Smooth manifolds

To define a Lie group we need the concept of a smooth manifold. A *manifold* is simply a topological space X which is locally homeomorphic to some Euclidean space $\mathbb{R}^n$, i.e. each point of X has a neighbourhood U which is homeomorphic to an open subset V of $\mathbb{R}^n$. Such homeomorphisms $\psi : U \to V$ are called *charts* for the manifold.

A *smooth manifold* is a manifold X together with a preferred collection of charts $\psi_\alpha : U_\alpha \to V_\alpha$ which cover all of X and are *smoothly related*, i.e. for any α, β the transition map $\psi_{\alpha\beta} = \psi_\beta \circ \psi_\alpha^{-1}$ from $\psi_\alpha(U_\alpha \cap U_\beta)$ to $\psi_\beta(U_\alpha \cap U_\beta)$ is smooth. (I shall use "smooth" to mean C^∞, i.e infinitely differentiable.) The preferred collection of charts is called the *atlas* of the manifold X, or simply the "smooth charts". It is best to assume that the atlas is *maximal*, i.e. that any chart which is smoothly related to all the charts of the atlas belongs to the atlas.

Example. One can cover the sphere

$$S^2 = \{(x,y,z) \in \mathbb{R}^3 : x^2 + y^2 + z^2 = 1\}$$

by six open sets $U_1, \ldots, U_6$, where

U_1 consists of the points where $x > 0$,
U_2 consists of the points where $x < 0$,
U_3 consists of the points where $y > 0$,

and so on. There are obvious charts $\psi_i : U_i \to V_i \subset \mathbb{R}^2$; for example, $\psi_1(x,y,z) = (y,z)$. These charts are smoothly related — e.g. the transition map ψ_{13} is given by

$$\psi_{13}(y,z) = (+(1-y^2-z^2)^{\frac{1}{2}}, z).$$

They define a smooth structure on S^2. Another chart belonging to the same atlas is the one given by stereographic projection (see page 55) from the north pole $N = (0, 0, 1)$ to the equatorial plane $z = 0$. This is the homeomorphism ψ from $U = S^2 - \{N\}$ to $\mathbb{R}^2$ defined by

$$\psi(x, y, z) = (x/(1-z), y/(1-z)).$$

At first sight one would say that a manifold is a very natural concept, while a smooth manifold seems a cumbersome and inconvenient thing. But that turns out to be quite wrong: experience shows that smooth manifolds are very practical, while manifolds in general are intractable. It is worth emphasizing that a smooth manifold is completely described by giving the set of points and prescribing which real-valued functions on it are smooth.

Manifolds sometimes arise embedded as submanifolds of Euclidean space, but sometimes they do not. The orthogonal group O_3 is naturally a 3-dimensional submanifold of the space $\mathbb{R}^9$ of all 3×3 matrices, defined by the six equations

$$O_3 = \{A : A^t A = 1\}.$$

(There are six equations here, because $A^t A$ is symmetric.) Charts for O_3 can be given in many ways. One elegant way is the *Cayley parametrization*: if

$$U = \{A \in O_3 : \det(A + 1) \neq 0\},$$

then a bijection

$$\psi : U \longrightarrow V = \{\text{skew } 3 \times 3 \text{ matrices}\} \cong \mathbb{R}^3$$

is defined by $\psi(A) = (A - 1)(A + 1)^{-1}$. The group is covered by the open sets $\{gU\}$ for all $g \in O_3$ – actually it is enough to let g run through the eight diagonal elements of the group — and a chart $\psi_g : gU \to V$ is defined by $\psi_g(A) = \psi(g^{-1}A)$.

A good example of a manifold which does not arise naturally as a subset of Euclidean space is the *projective space* $\mathbb{P}_\mathbb{R}^{n-1} = \mathbb{P}(\mathbb{R}^n)$, which consists of all lines through the origin in $\mathbb{R}^n$. A point of $\mathbb{P}_\mathbb{R}^{n-1}$ is represented by n *homogeneous coordinates* $(x_1, \ldots, x_n)$, not all zero, and $(x_1, \ldots, x_n)$ represents the same point as $(\lambda x_1, \ldots, \lambda x_n)$ if $\lambda \neq 0$. If U_n is the part of $\mathbb{P}_{\mathbb{R}}^{n-1}$ consisting of points with $x_n \neq 0$ then we have a bijection $\psi_n : U_n \to \mathbb{R}^{n-1}$ given by

$$\psi_n(x_1,\ldots,x_n) = (x_1 x_n^{-1},\ldots,x_{n-1} x_n^{-1}).$$

Obviously $\mathbb{P}_{\mathbb{R}}^{n-1}$ is covered by n such sets $U_1,\ldots,U_n$, with bijections $\psi_i : U_i \to \mathbb{R}^{n-1}$. One readily checks that they define a smooth atlas. Notice that in situations like this we do not need to define a topology on $\mathbb{P}_{\mathbb{R}}^{n-1}$ explicitly: the atlas provides it with a topology which makes it a manifold.

Only slightly more general is the case of the *Grassmannian* $\mathrm{Gr}_k(\mathbb{R}^n)$, which is the set of all k-dimensional vector subspaces of $\mathbb{R}^n$. A point W of $\mathrm{Gr}_k(\mathbb{R}^n)$ is represented by an $n \times k$ matrix x of rank k, whose columns form a basis for W. In this case x and $x\lambda$ represent the same point if λ is an invertible $k \times k$ matrix. For each k-element subset S of $\{1,\ldots,n\}$ let U_S be the part of $\mathrm{Gr}_k(\mathbb{R}^n)$ represented by matrices x whose S^{th} $k \times k$ submatrix x_S is invertible. As with projective space, $x \mapsto x x_S^{-1}$ defines a bijection between U_S and the vector space of $(n-k) \times k$ matrices. (For $x x_S^{-1}$ is an $n \times k$ matrix whose S^{th} block is the $k \times k$ identity matrix.) The reader may like to check that the transition map between U_S and U_T is

$$x \longmapsto (a + bx)(c + dx)^{-1},$$

where

$$\begin{pmatrix} a & b \\ c & d \end{pmatrix}$$

is the permutation matrix corresponding to the shuffle which takes S to T.

If X and Y are smooth manifolds then by using the charts we can say what it means for a map $f : X \to Y$ to be smooth. To be precise, f is smooth if $\tilde{\psi} \circ f \circ \psi^{-1}$ is a smooth map from $\psi(U \cap f^{-1}\tilde{U})$ to $\tilde{V}$ whenever $\psi : U \to V$ and $\tilde{\psi} : \tilde{U} \to \tilde{V}$ are charts for X and Y.

We can now give the long-postponed definition of a Lie group.

Definition 5.1

A Lie group is a smooth manifold G together with a smooth map $G \times G \to G$ which makes it a group.

Any closed subgroup of $GL_n\mathbb{R}$ is a Lie group, but I shall omit the proof. (A short elegant proof can be found in [Adams], pages 17–19.)

By applying the implicit function theorem to solve the equation $xy = 1$ for y in terms of x one finds that in any Lie group the map $x \mapsto x^{-1}$ is a smooth map $G \to G$.

Finally, the smoothness requirement in the definition of a Lie group is actually superfluous. But that is difficult and tedious to prove, and, as far as I know, it is a theorem without any applications. (It was proposed by Hilbert in 1900 as the fifth of his celebrated problems for the 20th century, and was proved by Gleason, Montgomery, and Zippin in 1953.) A closely related fact, but much more useful and quite easy to prove, is that any continuous homomorphism of Lie groups is smooth.

Tangent spaces

A smooth n-dimensional manifold X has a *tangent space* T_xX at each point x. It is an n-dimensional real vector space.

If X is a submanifold of $\mathbb{R}^N$ one can think of T_xX as a vector subspace of $\mathbb{R}^N$. We consider all smooth curves $\gamma : (-\epsilon, \epsilon) \to X$ such that $\gamma(0) = x$. Then T_xX is the set of all the velocity vectors $\gamma'(0) \in \mathbb{R}^N$. Equivalently, if $\phi : V \to X \subset \mathbb{R}^N$ is a local parametrization such that $\phi(y) = x$, then T_xX is the image of the linear map $D\phi(y) : \mathbb{R}^n \to \mathbb{R}^N$. (Here a *local parametrization* $\phi : V \to X$ means a map which is the inverse of a chart $\psi : U \to V$, where V is an open subset of $\mathbb{R}^n$.)

But we can define T_xX without invoking the ambient space $\mathbb{R}^N$: an element of T_xX is defined by a triple (x, ψ, ξ), where $\psi : U \to V$ is a chart such that $x \in U$, and ξ is a vector in $\mathbb{R}^n$ which we think of as the representative of the element of T_xX with respect to the chart ψ. A triple (x, ψ, ξ) is regarded as defining the *same* tangent vector as $(x, \tilde{\psi}, \tilde{\xi})$ if and only if $\tilde{\xi} = D\theta(y)\xi$, where $\theta = \tilde{\psi} \circ \psi^{-1}$ in a neighbourhood of y.

Example. If $G = O_n$, regarded as a submanifold of the $n \times n$ matrices, then T_1G is the $\frac{1}{2}n(n-1)$-dimensional vector space S of all skew matrices, and $T_gG = gS = Sg$.

Proof. For any skew matrix A the matrix e^{tA} is orthogonal, so $\gamma(t) = ge^{tA}$ defines a path $\gamma : \mathbb{R} \to G$ such that $\gamma(0) = g$ and $\gamma'(0) = gA$. Conversely, if $\gamma : (-\epsilon, \epsilon) \to G$ is a path such that $\gamma(0) = g$ then by differentiating $\gamma^t\gamma = 1$ we find

$$\gamma'(0)^t g + g^t \gamma'(0) = 0,$$

which shows that $g^{-1}\gamma'(0)$ is skew, i.e. that $T_gG \subset gS$.

5 Lie theory

Exercise. If $G = U_n$ then $T_1 G$ is the n^2-dimensional real vector space of skew hermitian matrices.

Notation

(i) A smooth map $f : X \to Y$ obviously induces a linear map $T_x X \to T_{f(x)} Y$ for any $x \in X$, and it is natural to denote this map by $Df(x)$.

(ii) If G is a Lie group, and $g \in G$, there is a smooth map $L_g : G \to G$ given by left-translation, i.e. $L_g(x) = gx$. This induces an isomorphism $T_x G \to T_{gx} G$ which I shall write simply as $\xi \mapsto g\xi$, thinking of the matrix-group example above. The corresponding isomorphism $T_x G \to T_{xg} G$ given by right-translation will be written $\xi \mapsto \xi g$.

One-parameter subgroups and the exponential map

A homomorphism $f : \mathbb{R} \to GL_n(\mathbb{R})$ – a so called *one-parameter subgroup* – is necessarily of the form $f(t) = e^{tA}$, where A is the matrix $f'(0)$. For

$$\begin{aligned} f'(t) &= \lim_{h \to 0} h^{-1}\{f(t+h) - f(t)\} \\ &= \lim_{h \to 0} h^{-1}\{f(h) - 1\}f(t) \\ &= Af(t), \end{aligned}$$

and the unique solution of the differential equation $f'(t) = Af(t)$ such that $f(0) = 1$ is $f(t) = e^{tA}$.

Furthermore, the map $\exp : M_n \mathbb{R} \to GL_n \mathbb{R}$ is bijective in a neighbourhood of zero, its inverse being the smooth map $g \mapsto \log g$ defined, when $\| g - 1 \| < 1$, by

$$\log(1 - A) = -\Sigma A^k / k.$$

Theorem 5.2 *In any Lie group G there is a 1-1 correspondence between the tangent space $T_1 G$ and the homomorphisms $f : \mathbb{R} \to G$.*

Proof. The argument is essentially the same as for $GL_n \mathbb{R}$. A homomorphism f gives us a tangent vector $f'(0) \in T_1 G$. Conversely, $A \in T_1 G$ defines a tangent vector field ξ_A on G by $\xi_A(g) = Ag$, and we have only to show that ξ_A has a unique solution curve with $f(0) = 1$.

The theory of differential equations gives us a solution

$$f : (-\epsilon, \epsilon) \to G$$

for some $\epsilon > 0$. It is a homomorphism where defined because both $t \mapsto f(t + u)$ and $t \mapsto f(t)f(u)$ are solution curves of ξ_A which take the value $f(u)$ when $t = 0$. But then for any $t \in \mathbb{R}$ the element $f(t/n)^n$ is defined for all sufficiently large n, and is independent of n, because

$$f(t/n)^n = f(t/nm)^{nm} = f(t/m)^m.$$

So we can define $f(t) = f(t/n)^n$ for any large n.

We have, therefore, a map

$$\exp : T_1 G \to G$$

whose derivative at 0 is the identity. In general it is neither 1-to-1 nor onto, but by the Inverse Function Theorem there is a smooth inverse map which we call 'log' defined in a neighbourhood of 1 in G.

Examples.

(i) If $G = SL_n\mathbb{R}$ then $T_1 G$ consists of the $n \times n$ matrices with trace 0, because $\det(e^{tA}) = e^{\text{trace}(tA)}$.

(ii) If $G = SU_2$ then $T_1 G$ is the skew-hermitian matrices with trace 0, i.e. the pure vector quaternions $\mathbb{R}^3$. If $u \in \mathbb{R}^3$ is a unit vector, regarded as a quaternion, then $u^2 = -1$, so

$$\exp(tu) = \cos t + u \sin t \ .$$

This is the 1-parameter subgroup of rotations about u. In particular, exp is surjective.

Remark. In fact exp is surjective in *any* compact group, for such a group has a Riemannian metric for which the geodesics emanating from 1 are precisely the 1-parameter subgroups, and in a complete Riemannian manifold any two points can be joined by a geodesic.

In non-compact groups exp is usually not surjective.

The next four examples can serve as exercises for the reader.

(iii) (See Chapter 2) In $SL_2\mathbb{R}$ the elements with trace > -2 are on 1-parameter subgroups. Those with trace ≤ -2 (i.e. those in the region C of the diagram on page 57) are not, with the exception of -1.

(iv) In $SL_2\mathbb{C}$ the matrix $\begin{pmatrix} -2 & 0 \\ 0 & -\frac{1}{2} \end{pmatrix}$ is on a 1-parameter subgroup, but *not* on any 1-parameter subgroup of $SL_2\mathbb{R}$.

(v) In $GL_2\mathbb{C}$ the matrix $\begin{pmatrix} -1 & 1 \\ 0 & -1 \end{pmatrix}$ is on a 1-parameter subgroup, but *not* on any 1-parameter subgroup of $SL_2\mathbb{C}$.

(vi) In $GL_n\mathbb{C}$ the map exp is surjective, as one can see by using the Jordan normal form.

5 Lie theory

Lie's theorems

In a Lie group G with $T_1 G = \mathfrak{g}$ the map

$$\log : U \to \mathfrak{g}$$

inverse to the exponential map is a canonical chart defined in a neighbourhood U of the identity element. It is natural to ask what the composition law $G \times G \to G$ looks like in this chart, i.e. how to express

$$C(A, B) = \log(\exp(A)\exp(B))$$

in terms of A and B. We can expand C in a Taylor series at $A = B = 0$:

$$C(A, B) = A + B + \tfrac{1}{2} b\,(A, B) + \text{(terms of order} \geq 3), \qquad (5.3)$$

where $\tfrac{1}{2} b\,(A, B)$ is the second order term. Because

$$C(A, 0) = A \text{ and } C(0, B) = B$$

the map b is necessarily a *bilinear* map

$$b : \mathfrak{g} \times \mathfrak{g} \longrightarrow \mathfrak{g},$$

and it is *skew* because $C(-B, -A) = -C(A, B)$.

One way of stating the basic miracle of Lie theory is that
(i) the infinite series (5.3) can be expressed entirely in terms of the bilinear map b, and
(ii) the series *converges* in a neighbourhood of the origin.
For example, the third order terms are

$$\tfrac{1}{12} b(A, b(A, B)) + \tfrac{1}{12} b(B, b(B, A)).$$

The complete series (5.3) is called the *Campbell-Baker-Hausdorff* series.

By direct calculation one finds that in a matrix group

$$b(A, B) = [A, B] = AB - BA,$$

so one writes $[\,,\,] : \mathfrak{g} \times \mathfrak{g} \longrightarrow \mathfrak{g}$ for the skew bilinear map b in general. It is easily seen to satisfy the Jacobi identity

$$[[A, B], C] + [[B, C], A] + [[C, A], B] = 0.$$

In other words, it makes $\mathfrak{g}$ into a *Lie algebra*.

Example. If $G = SO_3$ then $\mathfrak{g}$ is the 3×3 real skew matrices, and can be identified with $\mathbb{R}^3$. The Lie bracket $\mathbb{R}^3 \times \mathbb{R}^3 \to \mathbb{R}^3$ is the "vector product" of elementary geometry. The Jacobi identity follows from the well-known formula

$$(a \times b) \times c = \langle a, c \rangle b - \langle b, c \rangle a.$$

But the miracle of Lie theory is even better than I have said. The picture which Lie worked out is stated in modern language as

Theorem 5.4 *The functor taking G to $T_1 G$ is an equivalence of categories between the category of connected simply connected Lie groups and the category of Lie algebras.*

This means that every Lie algebra $\mathfrak{g}$ arises from a simply connected Lie group G, and that G is determined up to isomorphism by $\mathfrak{g}$. Futhermore, group homomorphisms $G_1 \to G_2$ are in 1-1 correspondence with Lie algebra homomorphisms $T_1 G_1 \to T_1 G_2$. The theorem reduces the study of Lie groups to questions in the vastly simpler realm of linear algebra.

After 120 years there is still no altogether easy proof of Lie's theorem. I shall try to sketch the main ideas in the remainder of this section. Three preliminary remarks may be helpful.

(i) Groups which are locally isomorphic clearly have the same Lie algebra. The theorem tells us that there is precisely one connected and simply connected group locally isomorphic to a given group. One half of this is easy: any group G is locally isomorphic to its simply connected covering group $\tilde{G}$, whose elements are pairs (g, γ), where $g \in G$ and γ is a homotopy class of paths in G from 1 to g.

When G is SO_3 the definition of $\tilde{G}$ is appealingly illustrated by the party trick called "Dirac's spanner". An element $g \in G$ is represented by a rigid body such as a spanner or undergraduate, whose centre of mass is fixed. The path γ is represented by a collection of strings which run from g's hands and feet to fixed points in space. Experiment shows that by rotating g one can get the strings γ into exactly two states — "tangled" and "untangled" — for each position of g, and that rotating g through 360° interchanges the two states of γ.

(ii) When a homomorphism $\theta : T_1 G_1 \to T_1 G_2$ of Lie algebras is given it is obvious that there is *at most* one group homomorphism $f : G_1 \to G_2$ which induces it. For f is determined by its restriction to a neighbourhood of 1, and hence by its values $f(\exp \xi)$ for $\xi \in T_1 G_1$. But

$f(\exp \xi) = \exp \theta(\xi)$, because $t \mapsto f(\exp t\xi)$ and $t \mapsto \exp t\theta(\xi)$ are both 1-parameter subgroups of G_2 with the same derivative at $t = 0$. One must remember, however, that $\exp : T_1 G \to G$ is usually not surjective. (See page 74.)

(iii) If $\mathfrak{h}$ is a sub-Lie-algebra of $T_1 G$ there is a Lie group H with $T_1 H = \mathfrak{h}$ and a homomorphism $H \to G$ inducing the inclusion $\mathfrak{h} \to T_1 G$. But the image of H in G need not be closed, and H need not be homeomorphic to any topological subgroup of G. The classical example is when G is a torus $\mathbb{T} \times \mathbb{T}$ and $\mathfrak{h} \subset \mathbb{R} \oplus \mathbb{R} = T_1 G$ is a line of irrational slope. If $\mathfrak{h}$ has rational slope p/q the corresponding 1-parameter subgroup $f : \mathbb{R} \to G$ closes up and forms a circle in the torus after winding p times round the left-hand $\mathbb{T}$ and q times round the right-hand $\mathbb{T}$. But when the slope is irrational $f : \mathbb{R} \to G$ is injective, and its image is a curve which winds densely round the torus.

Turning now to the proof of Lie's theorem, I think the best place to start is with the problem of constructing a homomorphism $f : G_1 \to G_2$ of Lie groups when one is given a homomorphism $\theta : \mathfrak{g}_1 \to \mathfrak{g}_2$ between their Lie algebras.

In a neighbourhood of the identity f can be defined by $f(\exp \xi) = \exp \theta(\xi)$. One way to prove the theorem is to show that this is a homomorphism (where it is defined) by constructing the Campbell–Baker–Hausdorff series (5.3) explicitly, and proving that it converges. This is arduous; but it is elegantly described in [Serre]. Even then one has still to extend f to the whole group, which involves using the simple-connectedness of G_1.

The method essentially used by Lie seems much more illuminating to me; in particular, it makes clear where the simple-connectedness is used.

To define $f(g)$ we choose a smooth path $\gamma : [0, 1] \to G$ from 1 to g. Then we consider the path

$$t \longmapsto \gamma'(t)\gamma(t)^{-1} = \xi(t) \text{, say,}$$

in the Lie algebra $\mathfrak{g}_1$. We transfer this to a path

$$t \longmapsto \theta(\xi(t)) = \tilde{\xi}(t) \text{, say,}$$

in $\mathfrak{g}_2$, then we solve† the ordinary differential equation

$$\varphi'(t) = \tilde{\xi}(t)\varphi(t)$$

in G_2, with initial condition $\varphi(0) = 1$. Finally, we define $f(g) = \varphi(1)$. The main point is to show that $\varphi(1)$ does not depend on the choice of the path γ from 1 to g.

If we have two different paths then, using the simple-connectedness, they form part of a family of paths

$$\{t \longmapsto \gamma_s(t)\}_{0 \leqslant s \leqslant 1},$$

all from 1 to g. Let

$$\xi(t,s) = \frac{\partial}{\partial t}\gamma_s(t).\gamma_s(t)^{-1} \in \mathfrak{g}_1,$$

$$\eta(t,s) = \frac{\partial}{\partial s}\gamma_s(t).\gamma_s(t)^{-1} \in \mathfrak{g}_1.$$

We calculate that

$$\frac{\partial \xi}{\partial s} - \frac{\partial \eta}{\partial t} = [\eta, \xi]. \tag{5.5}$$

This is called the *Maurer-Cartan* equation. Define $\tilde{\xi} = \theta \circ \xi$ and $\tilde{\eta} = \theta \circ \eta$, so that – because θ is a homomorphism of Lie algebras –

$$\frac{\partial \tilde{\xi}}{\partial s} - \frac{\partial \tilde{\eta}}{\partial t} = [\tilde{\eta}, \tilde{\xi}]. \tag{5.6}$$

This is precisely the compatibility condition which enables us to solve‡ the pair of equations

$$\frac{\partial \theta}{\partial t} = \tilde{\xi}\varphi, \quad \frac{\partial \theta}{\partial s} = \tilde{\eta}\varphi \tag{5.7}$$

to obtain $\varphi : [0, 1] \times [0, 1] \to G$. (In modern language, the Maurer-Cartan equations (5.5) and (5.6) express the fact that the Lie-algebra-valued 1-forms $A = \xi dt + \eta ds$ and $\tilde{A} = \tilde{\xi} dt + \tilde{\eta} ds$ are *flat connections* on $\mathbb{R}^2$. The equation (5.5) can be written $dA = \frac{1}{2}[A, A]$, and (5.6) is $d\tilde{A} = \frac{1}{2}[\tilde{A}, \tilde{A}]$.)

---

† If one thinks of γ as the path of an aircraft flying through G_1, then ξ is the record in its flight recorder, and φ is the path of an aircraft which flies in G_2 according to the programme $\tilde{\xi} = \theta \circ \xi$.

‡ If (5.7) holds then (5.6) follows by equating $\partial/\partial s(\partial \varphi/\partial t)$ to $\partial/\partial t(\partial \varphi/\partial s)$. Conversely, if (5.6) holds one can first define $\varphi(t, 0)$ by integrating $\partial \varphi/\partial t = \tilde{\xi}\varphi$ along the line $s = 0$, and then define $\varphi(t, s)$ by integrating $\partial \varphi/\partial s = \tilde{\xi}\varphi$ holding t constant. Then (5.6) tells us that $\partial/\partial s\{\partial \varphi/\partial t - \tilde{\xi}\varphi\} = 0$, which implies that φ satisfies both equations (5.7).

Now η and $\tilde{\eta}$ vanish when $t = 1$ by definition, so $\partial\varphi/\partial s = 0$ when $t = 1$, and $\varphi(1, s)$ is independent of s, as we want.

The most difficult part of Lie's theorem is the proof that any (finite dimensional) Lie algebra arises from a Lie group. The easiest route is first to prove Ado's theorem that $\mathfrak{g}$ is isomorphic to a subalgebra of the Lie algebra of matrices $M_n\mathbb{R}$. Then we consider all smooth maps

$$\xi : [0, 1] \longrightarrow \mathfrak{g} \subset M_n\mathbb{R}$$

which vanish together with all their derivatives at 0 and 1. For each such ξ we solve the differential equation

$$\varphi'_\xi(t) = \xi(t)\varphi_\xi(t) \tag{5.8}$$

to obtain $\varphi_\xi : [0, 1] \to GL_n\mathbb{R}$ such that $\varphi_\xi(0) = 1$. The elements $\varphi_\xi(1)$ of $GL_n\mathbb{R}$ obtained by this process form a subgroup of $GL_n\mathbb{R}$, for

$$\varphi_\eta(1)\varphi_\xi(1) = \varphi_{\eta*\xi}(1), \tag{5.9}$$

where $\eta * \xi : [0, 1] \to \mathfrak{g}$ is the *concatenation* of ξ and η, i.e.

$$(\eta * \xi)(t) = \begin{cases} 2\xi(2t) & \text{if } 0 \leq t \leq \frac{1}{2} \\ 2\eta(2t - 1) & \text{if } \frac{1}{2} \leq t \leq 1. \end{cases}$$

(To prove (5.9) we observe that, if

$$\varphi(t) = \begin{cases} \varphi_\xi(2t) & \text{for } 0 \leq t \leq \frac{1}{2} \\ \varphi_\eta(2t - 1)\varphi_\xi(1) & \text{for } \frac{1}{2} \leq t \leq 1, \end{cases}$$

then φ satisfies $\varphi' = (\eta * \xi)\varphi$.)

The subgroup of $GL_n\mathbb{R}$ so defined is almost, but not quite, the group we want to associate to $\mathfrak{g}$: unfortunately it may not be a closed subgroup of $GL_n\mathbb{R}$, as the dense winding on the torus illustrated. Instead, we consider the vector space $\mathscr{P}$ of all maps ξ as above, and introduce the equivalence relation

$$\xi_0 \sim \xi_1 \iff \varphi_{\xi_0}(1) = \varphi_{\xi_1}(1). \tag{5.10}$$

The quotient space $\mathscr{P}/\sim$ is a topological group under the operation of concatenation. It is the group we want. We must show it is locally homeomorphic to the Lie algebra $\mathfrak{g}$. But if $\xi \in \mathscr{P}$ is small then $\hat{\varphi}_\xi = \log\varphi_\xi$ is a well-defined path in $M_n\mathbb{R}$. We shall show that it is actually contained in $\mathfrak{g}$. This means that locally $\mathscr{P}/\sim$ is the same as the space of smooth paths $\hat{\varphi} : [0, 1] \to \mathfrak{g}$ with $\hat{\varphi}(0) = 1$, modulo the equivalence relation

$$\hat{\varphi}_0 \sim \hat{\varphi}_1 \iff \hat{\varphi}_0(1) = \hat{\varphi}_1(1).$$

In other words, locally $\mathcal{P}/\sim$ looks like $\mathfrak{g}$. One must check that the composition law in $\mathcal{P}/\sim$ is smooth, but that is easy.

It remains to give the proof that the path $\hat{\varphi}_\xi$ lies in $\mathfrak{g}$. Its velocity $\hat{\varphi}'_\xi(t)$ is related by the derivative of the exponential map to $\varphi'_\xi(t)$, and $\varphi'_\xi(t)\varphi_\xi(t)^{-1}$ belongs to $\mathfrak{g}$. There is an elegant formula for the derivative of the exponential map† :

$$\delta(e^A)e^{-A} = F(\text{ad } A)\delta A, \tag{5.11}$$

where $F : \text{End}(M_n\mathbb{R}) \to \text{End}(M_n\mathbb{R})$ is defined by

$$F(x) = (e^x - 1)/x = \sum_{k \geq 0} x^k/(k+1)!.$$

Here ad $A \in \text{End}(M_n\mathbb{R})$ is given by ad $A(B) = [A, B]$. The formula (5.11) shows that $A(t) = \hat{\varphi}_\xi(t)$ satisfies the differential equation

$$F(\text{ad } A)A' = \xi(t) \tag{5.12}$$

for a function $A : [0, 1] \to \mathfrak{g}$, and this completes the proof. I should, however, say a word about the derivation of (5.11). It is got by combining two results whose proofs can be left as straightforward exercises, namely

$$\frac{d}{dt}(e^A) = \int_0^1 e^{sA}\frac{dA}{dt}e^{(1-s)A}ds$$

for any smooth function $A : \mathbb{R} \to M_n\mathbb{R}$, and

$$e^{\text{ad } A}(B) = e^A B e^{-A}.$$

Finally, it is easy to see how we could have avoided invoking Ado's theorem. It was used only to define the equivalence relation (5.10) on $\mathcal{P}$. We could have replaced that by prescribing $\xi_0 \sim \xi_1$ if ξ_0 and ξ_1 are joined by a path $\{\xi_s\}$ in $\mathcal{P}$ such that the Maurer-Cartan equation (5.5) is satisfied for some $\eta : [0, 1] \times [0, 1] \to \mathfrak{g}$. This gives us a topological group $\mathcal{P}/\sim$ as before. To prove it is locally like $\mathfrak{g}$ we associate to each small

† It holds in any Lie group.

5 Lie theory

$\xi \in \mathscr{P}$ a path $\hat{\varphi}_\xi$ in $\mathfrak{g}$ by solving the equation (5.12). The hardest step is to check that

$$\xi_0 \sim \xi_1 \iff \hat{\varphi}_{\xi_0}(1) = \hat{\varphi}_{\xi_1}(1).$$

We shall not do that here.

The discussion of Lie's theorem I have given has the advantage that it can mostly be applied to infinite dimensional groups, though I should emphasize that the theorem is not true without restrictions for infinite dimensional groups. A very attractive account of the subject can be found in [Milnor][3], which influenced my treatment here.

6
Fourier series and representation theory

From now on these lectures will mostly be concerned with *representations* of Lie groups G. This means that we have a topological vector space V, and

(i) each $g \in G$ defines a linear isomorphism $V \to V$, written $v \mapsto gv$,
(ii) $(g_1 g_2)v = g_1(g_2 v)$ for $g_1, g_2 \in G$ and $v \in V$, and
(iii) $(g, v) \mapsto gv$ is a continuous map $G \times V \to V$.

We shall always assume that the vector space V is *complex*, and also locally convex and complete. The significance of the last two conditions is that they permit us to integrate any continuous function $f : [a, b] \to V$ to obtain an element $\int_a^b f(x)dx$ of V.

One of the most important theorems is mathematics is Fourier's theorem, which asserts that a smooth function $f : \mathbb{T} \to \mathbb{C}$ on the circle can be expanded in a Fourier series

$$f(\theta) = \sum_{n \in \mathbb{Z}} a_n e^{in\theta}, \quad \text{where } a_n = \int_{\mathbb{T}} f(\theta) e^{-in\theta} \frac{d\theta}{2\pi}. \tag{6.1}$$

This can be viewed as a theorem about group representations. If V is a representation – perhaps infinite-dimensional – of the group $\mathbb{T}$, then any $\xi \in V$ can be expanded

$$\xi = \sum_{n \in \mathbb{Z}} \xi_n, \quad \text{where } \xi_n = \int_{\mathbb{T}} (R_\theta \xi) e^{in\theta} \frac{d\theta}{2\pi}, \tag{6.2}$$

and

$$R_\theta \xi_n = e^{-in\theta} \xi_n. \tag{6.3}$$

Here $R_\theta : V \to V$ denotes the action of the group-element θ on V.

6 Representation theory

The proof of (6.2) is exactly the same as that of (6.1): substituting the definition of a_n or ξ_n into the corresponding series one sees that each result is equivalent to the fact that the functions s_N defined by

$$s_N(\theta) = \frac{1}{2\pi} \sum_{|n| \leq N} e^{in\theta}$$

tend to the delta-function $\delta(\theta)$ as $N \to \infty$. Furthermore, (6.2) really does contain (6.1). For if V is the space $C^\infty(\mathbb{T})$ of all smooth functions $\mathbb{T} \to \mathbb{C}$ then (6.3) implies that

$$(R_\alpha \xi_n)(\theta) = \xi_n(\theta - \alpha) = e^{-in\alpha} \xi_n(\theta) ,$$

and hence that $\xi_n(\alpha) = a_n e^{in\alpha}$, where $a_n = \xi_n(0)$.

We can state the result as

Theorem 6.4 *If V is a representation of $\mathbb{T}$ then $V = \widehat{\bigoplus_{n \in \mathbb{Z}}} V_n$, where*

$$V_n = \{\xi \in V : R_\alpha \xi = e^{-in\alpha}\xi \text{ for all } \alpha \in \mathbb{T}\} .$$

The notation $\widehat{\bigoplus}$ is meant to imply that each V_n is a closed subspace of V, and that each $\xi \in V$ has a *unique* convergent expansion $\xi = \sum \xi_n$ with $\xi_n \in V_n$. In other words, V is a completion of the algebraic direct sum $\bigoplus V_n$. But there are many possible completions: V cannot be reconstructed from the V_n without more information.

As a simple application of Fourier's theorem let us prove

Theorem 6.5 *The Heisenberg group N/Z of page 50 is not a matrix group.*

Proof. N/Z has a circle subgroup $\mathbb{T}$ formed by the matrices

$$g_t = \begin{pmatrix} 1 & 0 & t \\ 0 & 1 & 0 \\ 0 & 0 & 1 \end{pmatrix}$$

for $t \in \mathbb{R}$. It is trivial to check that

(i) $\mathbb{T}$ is contained in the centre of N/Z, and
(ii) each element of $\mathbb{T}$ is a commutator $uvu^{-1}v^{-1}$ in N/Z.

We must show that whenever we have a finite dimensional representation V of N/Z, i.e. a homomorphism

$$\rho : N/Z \to \text{Aut}(V)$$

into the group of automorphisms of V, then ρ is not injective. But V can be decomposed $V = \bigoplus V_n$ under the action of the subgroup $\mathbb{T}$. As $\mathbb{T}$ is in the centre of N/Z each V_n is an invariant subspace for N/Z. Now $g_t \in \mathbb{T}$ acts on V_n by multiplication by $e^{-2\pi i n t}$. But because it is a commutator it acts with determinant 1. This is a contradiction unless $n = 0$. So $V = V_0$, and $\mathbb{T}$ acts trivially on V, and ρ is not injective.

General remarks about representations

A representation V is called *irreducible* if it has no closed G-invariant subspaces, except 0 and V.

When a representation V is reducible, with a closed invariant subspace W, we can ask whether it *decomposes*, i.e. whether we can find an invariant subspace W' such that $V = W \oplus W'$. In general we cannot. For example, the group of matrices of the form $\begin{pmatrix} 1 & * \\ 0 & 1 \end{pmatrix}$ acts on $\mathbb{C}^2$, and the subspace $W = \begin{pmatrix} * \\ 0 \end{pmatrix}$ is invariant, but there is no other invariant subspace.

But if the representation V is *unitary*, i.e. if V is a Hilbert space and the inner product $\langle \, , \, \rangle$ is invariant, i.e.

$$\langle g\xi, g\eta \rangle = \langle \xi, \eta \rangle$$

for all $g \in G$ and $\xi, \eta \in V$, then W is invariant if and only if $W^\perp$ is invariant, and $V = W \oplus W^\perp$. In particular, a finite dimensional unitary representation is always a direct sum of irreducibles.

We shall constantly make use of a trivial but crucial remark about representations, which is always called **Schur's lemma**.

Lemma 6.6 *If V_1 and V_2 are finite dimensional irreducible representations of G then any G-map $f : V_1 \to V_2$ (i.e. linear map such that $f(g\xi) = gf(\xi)$) is either zero or an isomorphism.*

Furthermore, if $V_1 = V_2$ then $f = \lambda 1$ for some $\lambda \in \mathbb{C}$.

The first half of the lemma holds because $\ker(f)$ and $\text{im}(f)$ are invariant subspaces, the second half because any eigenspace of f is an invariant subspace.

7
Compact groups and integration

We should like to generalize the Fourier theory from $\mathbb{T}$ to other Lie groups. The only tool used was the ability to average functions over the group $\mathbb{T}$. The same can be done for any compact group.

If G is a compact group, V is a topological vector space (locally convex and complete), and $f : G \to V$ is continuous, then we can define

$$\int_G f(g)dg = \int_G f \in V.$$

More precisely, there is a continuous linear map

$$\int_G : C(G; V) \to V,$$

where $C(G; V) = \{\text{continuous maps } G \to V\}$, such that
(i) $\int_G f(g)dg = v$ if $f(g) = v$ for all g, and
(ii) $\int_G f(hg)dg = \int_G f(gh)dg = \int_G f(g)dg$ for any $h \in G$.

In addition, the measure on G is *positive*. This means that (i) can be generalized to

(iii) if f takes its values in a convex subset C of V, then $\int_G f \in C$.

From the general theory of integration on an orientable manifold we know that to define the integral it is enough to give a volume element at each point g of G, i.e. an alternating n-fold multilinear form ω_g on the tangent space $T_g G$, where $n = \dim(G)$. We can choose ω_1 arbitrarily, and then define ω_g at other points g by identifying $T_g G$ with $T_1 G$ by left-translation. This gives us a left-invariant integral $\int$, obviously unique up to a scalar multiple, which can be chosen so that $\int 1 = 1$. Performing a right-translation by h could only multiply the linear map $\int$ by a positive real number $\mu(h)$, which would give us a homomorphism $\mu : G \to \mathbb{R}_+^\times$. But if G is compact μ must be trivial, for $\mathbb{R}_+^\times$ has no compact subgroups. So the integral is right-invariant as well.

If $G = SU_2$, regarded as the unit sphere S^3, then $\int_G$ has the obvious meaning. It may be helpful to mention that if $G = GL_n\mathbb{R}$, which is not compact, the formula

$$\int_G f = \int f(A)\det(A)^{-n} dA,$$

where dA denotes ordinary Lebesgue measure on the vector space $\mathbb{R}^{n^2}$ of matrices, defines a left- and right-invariant integral for functions f with compact support on G; but of course then the function $f = 1$ is not integrable.

A formula for integration on U_n

It is not practical to give an explicit formula for integrating a general function on a group such as U_n, for there are no convenient coordinates to use. But if the function f to be integrated is a class-function, i.e. if $f(gxg^{-1}) = f(x)$ for all g and x, then there is a very elegant result. Its form and its derivation illustrate some aspects of Lie group theory so well that I shall include it, even though I shall make no use of it.

A class-function on U_n is a symmetric function of the eigenvalues $(u_1,\ldots,u_n)$ of the group element.

Theorem 7.1 *If $f : U_n \to \mathbb{C}$ is a class-function, then*

$$\int_{U_n} f = \frac{1}{n!} \int_0^{2\pi} \cdots \int_0^{2\pi} f(u_1,\ldots,u_n) \prod_{i<j} |u_i - u_j|^2 \frac{d\theta_1}{2\pi} \cdots \frac{d\theta_n}{2\pi},$$

where $u_k = e^{i\theta_k}$.

The formula tells us, essentially, that the volume of the $n(n-1)$-dimensional conjugacy class of the diagonal matrix with entries $(u_1,\ldots,u_n)$ is

$$\prod_{i<j} |u_i - u_j|^2.$$

(This vanishes if two of the u_i are equal, which is right, since then the conjugacy class has dimension less than $n(n-1)$.) The $1/n!$ in the formula corresponds to the fact that each conjugacy class is counted $n!$ times on

7 Compact groups and integration

the right-hand side: it could be removed if one restricted the integration to the region

$$0 \leqslant \theta_1 \leqslant \theta_2 \leqslant \cdots \leqslant \theta_n \leqslant 2\pi.$$

Proof. Let $G = U_n$, and let T denote the diagonal matrices. We consider the map

$$\begin{array}{rcl} T \times (G/T) & \to & G \\ (t, gT) & \mapsto & gtg^{-1} . \end{array}$$

Let the Jacobian of this map at (t, gT), with respect to invariant measures on T, G/T, and G, be $J(t)$: it is clear that it depends only on t and not on gT. Then for any function f on G we have

$$\int_G f = \frac{1}{n!} \int_{T \times (G/T)} f(gtg^{-1}) J(t) dt d(gT),$$

so if f is a class-function then

$$\int_G f = K^{-1} \int_T f(t) J(t) dt,$$

where $K^{-1} = \text{volume } (G/T)/n!$.

To calculate $J(t)$, let $\mathfrak{t}$ and $\mathfrak{g}$ denote the Lie algebras of T and G. We can identify the tangent space to G/T at its base-point with $\mathfrak{t}^\perp$, the orthogonal complement of $\mathfrak{t}$ in $\mathfrak{g}$. If t changes infinitesimally to $t(1 + \xi)$, with $\xi \in \mathfrak{t}$, and g changes from 1 to $1 + \eta$, where $\eta \in \mathfrak{t}^\perp$, then the change in gtg^{-1}, to first order, is

$$\begin{aligned} \rho &= (1+\eta)t(1+\xi)(1-\eta) - t \\ &= t\xi + \eta t - t\eta . \end{aligned}$$

So

$$t^{-1}\rho = \xi + (t^{-1}\eta t - \eta) \in \mathfrak{t} \oplus \mathfrak{t}^\perp = \mathfrak{g}.$$

and

$$J(t) = \det(A(t^{-1}) - 1),$$

where $A(t^{-1})$ denotes the adjoint action of t^{-1} on $\mathfrak{t}^\perp$. The complexification of $\mathfrak{t}^\perp$ has a basis consisting of the matrices E_{jk} for $j \neq k$, where E_{jk} has 1 in the (j,k) place and 0 elsewhere. Evidently E_{jk} is an eigenvector of $A(t^{-1})$ with eigenvalue $u_k u_j^{-1}$, if t has diagonal elements $(u_1, \ldots, u_n)$. So

$$J(t) = \prod_{j \neq k}(u_k u_j^{-1} - 1) = \prod_{j < k} |u_j - u_k|^2 .$$

We have now proved (7.1) apart from the calculation of the constant $K = \int_T J(t)dt$. But $J(t) = |\Delta(t)|^2$, where

$$\Delta(t) = \prod_{j<k}(u_j - u_k)$$

is the Vandermonde determinant

$$\begin{vmatrix} 1 & 1 & \cdots & 1 \\ u_1 & u_2 & \cdots & u_n \\ u_1^2 & u_2^2 & \cdots & u_n^2 \\ \vdots & \vdots & & \\ u_1^{n-1} & u_2^{n-1} & \cdots & u_n^{n-1} \end{vmatrix}.$$

Expanding the determinant

$$\Delta(t) = \sum \pm u_1^{m_1} u_2^{m_2} \cdots u_n^{m_n},$$

where $(m_1,\ldots,m_n)$ runs through the permutations of $\{0,1,\ldots,n-1\}$, and integrating each of the $(n!)^2$ terms separately, we find that $K = (2\pi)^n n!$.

The preceding calculation has been arranged so that it applies essentially without change to any compact Lie group G with maximal torus T. The matrices E_{jk} are replaced in general by the root vectors, and $J(t)$ becomes, in the notation of page 30, the function

$$\prod_{\alpha}(e(\alpha) - 1)$$

on T, where α runs through the roots of G. As before we have $J(t) = |\Delta(t)|^2$, where

$$\Delta(t) = \prod_{\alpha>0}(e(\alpha)^{1/2} - e(\alpha)^{-1/2}) = \sum_{w \in W} \det w \; e(w(\rho))$$

is the denominator of the Weyl character formula. The equality of the last two expressions, which is often called the "Weyl denominator formula", generalizes the product formula for the Vandermonde determinant.

8
Maximal compact subgroups

In the following sections I shall discuss the representation theory of compact groups in some detail. One of the main reasons for being interested in it is to use it as a tool for studying non-compact groups. The following basic structural theorem is therefore important.

Theorem 8.1 *In a Lie group G with a finite number of connected components there always exist maximal compact subgroups. If K is one of them, then any compact subgroup of G is conjugate to a subgroup of K, and in particular any two maximal compact subgroups are conjugate. Furthermore, G is homeomorphic to $K \times \mathbb{R}^m$ for some m.*

I shall prove this only for $GL_n\mathbb{R}$. But first let us consider the five 3-dimensional groups described in Chapter 1:

$$E_2, \quad SL_2\mathbb{R}, \quad SU_2, \quad N, \quad N/Z \ .$$

(i) The group E_2 acts on $\mathbb{R}^2$, and a compact subgroup H must leave a point fixed, for

$$\xi_0 = \int_H h\xi dh$$

is fixed under H for any $\xi \in \mathbb{R}^2$. So H is contained in the isotropy group G_{ξ_0} of some point ξ_0 of $\mathbb{R}^2$, and G_{ξ_0} is conjugate to the isotropy group O_2 of the origin. So O_2 is a maximal compact subgroup, and E_2 is clearly homeomorphic to $O_2 \times \mathbb{R}^2$.

(ii) $SL_2\mathbb{R}$ acts on the upper half-plane H, which is homeomorphic to $\mathbb{R}^2$. Again, though it is not quite so obvious, a compact subgroup must leave fixed some point of H, and so the maximal compact subgroups are the isotropy groups, such as the isotropy group SO_2 of $i \in H$. We saw in Chapter 4A that $SL_2\mathbb{R}$ is homeomorphic to $SO_2 \times \mathbb{R}^2$.

(iii) SU_2 is compact.

(iv) N is homeomorphic to $\mathbb{R}^3$, and its maximal compact subgroup is $\{1\}$.

(v) N/Z has the circle $\mathbb{T}$ of elements g_t (see page 83) as its unique maximal compact subgroup. It is a normal subgroup, and N/Z is homeomorphic to $\mathbb{T} \times \mathbb{R}^2$.

Now let us return to $GL_n\mathbb{R}$. To prove that any compact subgroup K is conjugate to a subgroup of O_n it is enough to prove that the action of K on $\mathbb{R}^n$ preserves an inner product on $\mathbb{R}^n$. For we can find an orthonormal basis with respect to any inner product by the Gram-Schmidt process (see Chapter 4B), and after changing to the orthornormal basis K will be represented by orthogonal matrices. We find a K-invariant inner product $\langle\langle\ ,\ \rangle\rangle$ by averaging the standard inner product $\langle\ ,\ \rangle$, i.e. we define

$$\langle\langle \xi, \eta \rangle\rangle = \int_K \langle k\xi, k\eta \rangle dk\ .$$

We saw in Chapter 4A that, as a topological space,

$$\begin{aligned} GL_n\mathbb{R} &\cong O_n \times \{\text{positive definite symmetric matrices}\} \\ &\cong O_n \times \mathbb{R}^{\frac{1}{2}n(n+1)}. \end{aligned}$$

A proof of Theorem (8.1) can be found in [Hochschild].

9
The Peter-Weyl theorem

We shall now describe the analogue of Fourier series for a compact Lie group G.

Let V be a representation of G. We can assume V is unitary, as we can find an invariant inner product by averaging an arbitary one, as was described at the end of Chapter 8.

If P is a finite dimensional irreducible representation of G we shall define a subspace V_P of V, called its *P-isotypical part*, consisting of the "vectors which transform according to P". This is analogous to the space

$$V_n = \{\xi \in V : R_\alpha \xi = e^{-in\alpha}\xi\}$$

in the Fourier case, for the irreducible representations of $\mathbb{T}$ are the 1-dimensional ones $\theta \mapsto e^{in\theta}$

We define V_P as the sum of all the copies of P contained in V. Equivalently, it is the image of the natural inclusion †

$$P \otimes \mathrm{Hom}_G(P;V) \longrightarrow V \qquad (9.1)$$

$$\xi \otimes f \longmapsto f(\xi),$$

where $\mathrm{Hom}_G(P;V)$ is the vector space of G-equivariant linear maps $P \to V$. One version of the Peter-Weyl theorem is

Theorem 9.2 *The isotypical part V_P is a closed subspace of V, and*

$$V = \widehat{\bigoplus_P} V_P,$$

---

† The map (9.1) is injective because it is injective when $V = P$ by Schur's lemma, and hence injective when V is a sum of copies of P. But the left-hand side of (9.1) does not change if V is replaced by the sum of all the copies of P contained in V.

where P runs through the finite dimensional irreducible **representations** of G.

The meaning of the notation $\hat{\oplus}$ is explained after Theorem (6.4).

Corollary 9.3 *All irreducible representations of G are finite dimensional.*

There are many equivalent but rather different-looking versions of the Peter-Weyl theorem. The second of the three I shall mention is

Theorem 9.4 *Any compact Lie group is isomorphic to a subgroup of a unitary group U_n. In particular, it is a matrix group.*

Of course it is a matter of tradition, perspective, and taste which theorems are regarded as "equivalent" to which, and it may seem odd to claim that Theorem (9.2) is equivalent to a theorem which is vacuous if G is U_n. The argument in favour of the tradition will be given presently.

To state the third version of the Peter-Weyl theorem we need a definition.

A *representative function* on G is a function $G \to \mathbb{C}$ of the form

$$f_{M;\xi,\eta}(g) = \langle \eta, g\,\xi \rangle,$$

where M is a finite dimensional unitary representation of G, and $\xi, \eta \in M$. (In other words, $f_{M;\xi,\eta}(g)$ is a matrix element of the action of g on M with respect to a suitable basis.) The representative functions form a subalgebra $C_{\text{alg}}(G)$ of the algebra $C(G)$ of continuous functions on G, for

$$f_{M_1;\xi_1,\eta_1} + f_{M_2;\xi_2,\eta_2} = f_{M_1 \oplus M_2;\xi_1 \oplus \xi_2,\eta_1 \oplus \eta_2},$$
$$f_{M_1;\xi_1,\eta_1} \cdot f_{M_2;\xi_2,\eta_2} = f_{M_1 \otimes M_2;\xi_1 \otimes \xi_2,\eta_1 \otimes \eta_2},$$

and

$$\lambda f_{M;\xi,\eta} = f_{M;\lambda\xi,\eta}.$$

In fact $C_{\text{alg}}(G)$ is the coordinate ring (in the sense of Macdonald's lectures) of a linear algebra group $G_\mathbb{C}$ which I shall return to in Chapter 12.

The third version of the Peter-Weyl theorem is

Theorem 9.5 $C_{\text{alg}}(G)$ *is a dense subring of $C(G)$ for the topology of uniform convergence.*

9 The Peter-Weyl theorem

Proof that (9.4) $\iff$ *(9.5).*

First suppose $G \subset M_n\mathbb{C}$. Then the matrix entries and their complex conjugates belong to $C_{\mathrm{alg}}(G)$. But by the Stone-Weierstrass theorem polynomials in these coordinate functions are dense in the continuous functions on the compact subset G of $M_n\mathbb{C}$.

Conversely, if $C_{\mathrm{alg}}(G)$ is dense in $C(G)$ then the representative functions separate the points of G, so any $g \in G$ acts non-trivially on *some* finite dimensional representation V_g of G. Let us consider the kernels of

$$G \longrightarrow \mathrm{Aut}(V_{g_1} \oplus \ldots, \oplus V_{g_n})$$

for larger and larger finite subsets $g_1, \ldots, g_n$ of G. Eventually the kernel must be $\{1\}$, for compact Lie groups obey the descending chain condition : in any decreasing chain either the dimension or the number of components must fall at each step.

Proof that (9.2) $\implies$ *(9.5).*

For any representation V of G we define the sub-vector-space V^{fin} of G-finite vectors by

$$V^{\mathrm{fin}} = \{\xi \in V : \xi \text{ is contained in some finite dimensional } G\text{-invariant subspace of } V\}.$$

Because any finite dimensional representation is a sum of irreducible representations Theorem (9.2) is equivalent to the assertion that V^{fin} is dense in V. So the implication we want follows from

Theorem 9.6 *If we let G act on $C(G)$ by left-translation, then*

$$C(G)^{\mathrm{fin}} = C_{\mathrm{alg}}(G).$$

Proof. If f belongs to a finite dimensional G-invariant subspace W of $C(G)$ then f is a matrix element of the action of G on $\bar{W}$. For if $\{f_1, \ldots, f_n\}$ is an orthonormal basis of W, with $f = f_1$, and g acts on W by

$$gf_i = \sum M_{ji}(g)f_j,$$

then

$$\begin{aligned} f(g) &= (g^{-1}f_1)(1) = \sum M_{j1}(g^{-1})f_j(1) \\ &= \sum \overline{M_{1j}(g)}f_j(1), \end{aligned}$$

so f is a linear combination of the matrix elements $\bar{M}_{1j}$.

Proof that (9.5) $\implies$ (9.2).

We must introduce a new construction. For each $f \in C(G)$ we can define an operator $T_f : V \to V$ as a weighted average of the action of the elements of G:

$$T_f \xi = \int_G f(g) g \xi \, dg.$$

(Physicists call this "smearing the G-action with f".) This makes V into a module over $C(G)$ when the usual product in $C(G)$ is replaced by the *convolution* product

$$(f_1 * f_2)(h) = \int_G f_1(g) f_2(g^{-1} h) \, dg.$$

If $f \in C_{\text{alg}}(G)$ then $T_f \xi \in V^{\text{fin}}$ for any $\xi \in V$, as one can easily check that $g T_f \xi = T_{gf} \xi$, and the functions gf, for all g, are linear combinations of the matrix elements of a single finite dimensional representation. There is no identity element in the convolution algebra $C(G)$: the identity element "would be" the delta-function at $1 \in G$, which is not continuous. But the delta-function can be approximated arbitrarily closely by continuous functions f satisfying $\int f = 1$ which are supported in a small neighbourhood of $1 \in G$. So for any $\xi \in V$ we can find $f \in C(G)$ such that $T_f \xi$ is very close to ξ. If (9.5) holds then we can find φ in C_{alg} very close to f. Thus $T_\varphi \xi$ is very close to ξ. But $T_\varphi \xi \in V^{\text{fin}}$, so V^{fin} is dense in V.

We have now shown that the various versions of the Peter-Weyl theorem are equivalent. If we are only interested in matrix groups there is no more to do. For general groups the most convenient version to prove is probably that $C(G)^{\text{fin}}$ is dense in $C(G)$. I shall give only a very brief sketch of the proof, referring to [Adams], or [Chevalley] for more details.

For any $f \in C(G)$ the operator $T_f : C(G) \to C(G)$ is an integral operator with a continuous kernel, for

$$\begin{aligned}(T_f \varphi)(x) &= \int_G f(g) \, \varphi(g^{-1} x) \, dg \\ &= \int_G f(xg^{-1}) \varphi(g) \, dg.\end{aligned}$$

Such an integral operator acting on the functions on a compact space is a compact operator. It is self-adjoint if f is *hermitian*, i.e. $\overline{f(g)} = f(g^{-1})$. The eigenspaces of a compact self-adjoint operator are finite dimensional, and any function in the range can be expanded in a uniformly convergent series of eigenfunctions. If we choose f to be conjugation-invariant (i.e. $f(xyx^{-1}) = f(y)$) then T_f commutes with the left-action of G on $C(G)$, and so the eigenspaces are finite dimensional subspaces, and the eigenfunctions belong to $C(G)^{\text{fin}}$. So if f is a conjugation-invariant hermitian approximation to the delta-function at $1 \in G$ its range is nearly the whole of $C(G)$, and $C_{\text{alg}}(G)$ is dense in its range. That completes the proof.

The structure of $C_{\text{alg}}(G)$

The group G acts on the representative functions $C_{\text{alg}}(G)$ by left-translation and also by right-translation, so altogther we have an action of $G \times G$ on it. The diagonal copy of G in $G \times G$ acts by conjugation. The whole situation can be described very elegantly.

Theorem 9.7 *(i) There is an isomorphism of representations of $G \times G$*

$$\bigoplus \overline{P} \otimes P \longrightarrow C_{\text{alg}}(G) \tag{9.8}$$

given by

$$\eta \otimes \xi \longmapsto f_{P;\eta,\xi},$$

where P runs through the irreducible unitary representations of G, and the left and right copies of G act on $\overline{P}$ and P respectively.†

(ii) Each $\overline{P} \otimes P$ is an irreducible representation of $G \times G$.

(iii) The isomorphism (9.8) takes the inner-product defined on the left as the orthogonal direct sum of the inner products

$$\langle \eta_1 \otimes \xi_1, \eta_2 \otimes \xi_2 \rangle = \frac{1}{\dim(P)} \overline{\langle \eta_1, \eta_2 \rangle} \langle \xi_1, \xi_2 \rangle$$

on $\overline{P} \otimes P$ to the usual L^2 inner product on $C_{\text{alg}}(G)$.

Before giving the proof of this theorem let us notice some of its many useful corollaries.

First, concerning characters. The *character* of a finite dimensional representation of V of G is the function $\chi_V : G \to \mathbb{C}$ defined by

$$\chi_V(g) = \text{trace}(g_V),$$

where $g_V : V \to V$ is the action of g on V. We have

† The complex conjugate representations $\overline{P}$ is identical with P as a set, but the scalar field $\mathbb{C}$ acts on it in the complex-conjugate way.

Corollary 9.9 *(i) A finite dimensional representation of G is determined up to isomorphism by its character.*

(ii) If P and Q are irreducible representations, then
$$\langle \chi_P, \chi_Q \rangle = \begin{cases} 1 & \text{if} \quad P \cong Q \\ 0 & \text{if not} \end{cases}$$

(iii) The characters of the irreducible representations form an orthonormal basis for the Hilbert space of class-functions *on G, i.e. functions f such that* $f(xyx^{-1}) = f(y)$.

To deduce the corollary from the theorem we notice that if $\{e_i\}$ is an orthonormal basis of an irreducible representation P then χ_P is the image of $\sum \bar{e}_i \otimes e_i$ under the map (9.8). So (9.9)(ii), and hence (9.9)(i), follows from (9.7)(iii). But (9.9)(iii) also follows, as the conjugation-invariant elements of $C_{\text{alg}}(G)$ correspond to the part of $\bigoplus \bar{P} \otimes P$ left fixed by the diagonal action of G, and this part is spanned by the single element $\sum \bar{e}_i \otimes e_i$, because

$$(\bar{P} \otimes P)^G \cong \text{Hom}_G(P; P) \cong \mathbb{C}$$

by Schur's lemma.

Another corollary, or really reformulation, whose justification I shall leave to the reader, is

Proposition 9.10 *The map*

$$C_{\text{alg}}(G) \longrightarrow \bigoplus \text{End}(P)$$

which takes φ to the collection of "smeared" operators $\varphi_P : P \to P$ is an isomorphism of algebras when $C_{\text{alg}}(G)$ is given the convolution product.

There is also an analytical aspect to Theorem (9.7). It tells us that any $f \in C_{\text{alg}}(G)$ can be expanded as a finite series

$$f = \sum f_P \qquad (9.11)$$

where f_P is in the image of $\bar{P} \otimes P$, and that

$$\| f \|^2 = \sum \frac{1}{\dim(P)} \| f_P \|^2, \qquad (9.12)$$

where $\| f \|$ is the L^2 norm, and $\| f_P \|$ is the natural norm on $\bar{P} \otimes P$. When functions on the circle are expanded as Fourier series

$$f(\theta) = \sum a_n e^{in\theta}$$

9 The Peter-Weyl theorem

we know that there are many correspondences of the type

L^2 functions f ⟷ square-summable sequences $\{a_n\}$,
C^∞ functions f ⟷ rapidly decreasing sequences $\{a_n\}$,
real-analytic functions f ⟷ exponentially decreasing sequences $\{a_n\}$.

Here "square-summable" means that $\sum |a_n|^2$ converges, "rapidly decreasing" means that $\{n^k a_n\}$ is bounded for all k, and "exponentially decreasing" means that $\{K^{|n|} a_n\}$ is bounded for some $K > 1$. These facts generalize to any compact group. The result for L^2 functions follows directly from (9.12). To state the others, which I shall not prove, one needs to know that the irreducible representations P of G are classified by their highest weights λ_P, and that λ_P has a norm $\| \lambda_P \|$. (See Chapter 14.)

Proposition 9.13 *Under the correspondence (9.11)*

L^2 functions f ⟷ square-summable sequences $\{f_P\}$,
C^∞ functions f ⟷ rapidly decreasing sequences $\{f_P\}$,
real-analytic functions f ⟷ exponentially decreasing sequences $\{f_P\}$.

In this case "square-summable" means that the right-hand side of (9.13) converges, "rapidly decreasing" means that $\{\| \lambda_P \|^k \| f_P \|\}$ is bounded for each k, and "exponentially decreasing" means that $\{K^{\|\lambda_P\|} \| f_P \|\}$ is bounded for some $K > 1$.

Finally, (9.7) gives us a description of the functions on a homogeneous space G/H, for $C(G/H)$ is just the part of $C(G)$ which is invariant under the right-hand action of H. We have

Corollary 9.14

$$C(G/H) \cong \bigoplus_P \hat{P} \otimes \overline{P}^H,$$

compatibly with the G-actions on $C(G/H)$ and the representations P.

Proof of (9.7). We begin with (ii). If P_1 and P_2 are irreducible representations of compact groups G_1 and G_2 then $P_1 \otimes P_2$ is an irreducible representation of $G_1 \times G_2$. To see this it is enough to show that

$$\operatorname{End}_{G_1 \times G_2}(P_1 \otimes P_2) \cong \mathbb{C}, \tag{9.15}$$

because representations of any compact group are sums of irreducibles. But the space of $(G_1 \times G_2)$-equivariant maps $P_1 \otimes P_2 \to P_1 \otimes P_2$ is the $(G_1 \times G_2)$-invariant part of the matrix algebra

$$\text{End }(P_1 \otimes P_2) \cong \text{End }(P_1) \otimes \text{End }(P_2),$$

and so (9.15) follows at once from Schur's lemma.

(i) When G acts on $C(G)$ by left-translation we know that $C(G)^{\text{fin}} = C_{\text{alg}}(G)$, and so that

$$C_{\text{alg}}(G) = \bigoplus \overline{P} \otimes \text{Hom }_G(\overline{P}; C(G)).$$

But there is a map

$$P \longrightarrow \text{Hom }_G(\overline{P}; C(G)), \tag{9.16}$$

equivariant with respect to the right-hand action of G on $C(G)$, defined by $\xi \mapsto F_\xi$, where

$$F_\xi(\eta)(g) = \langle \eta, g\xi \rangle.$$

The map (9.16) is injective by Schur's lemma, and is surjective because if $f : \overline{P} \to C(G)$ we can find $\xi \in P$ such that

$$\langle \eta, \xi \rangle = f(\eta)(1)$$

for all $\eta \in P$, and then $F_\xi = f$ because

$$F_\xi(\eta)(g) = \langle \eta, g\xi \rangle = \langle g^{-1}\eta, \xi \rangle = f(g^{-1}\eta)(1) = g^{-1}f(\eta)(1) = f(\eta)(g).$$

(iii) We first observe that the L^2 inner product on $C(G)$ induces a $(G \times G)$-invariant inner product on $\bigoplus \overline{P} \otimes P$. As the spaces $\overline{P} \otimes P$ are non-isomorphic irreducible representations of $G \times G$, Schur's lemma tells us that the summands $\overline{P} \otimes P$ are orthogonal for any invariant inner product, and that, up to a scalar multiple, there is only one invariant inner product on each $\overline{P} \otimes P$. We therefore have

$$\langle f_{P;\eta_1,\xi_1}, f_{P;\eta_2,\xi_2} \rangle = K_P \overline{\langle \eta_1, \eta_2 \rangle} \langle \xi_1, \xi_2 \rangle$$

for some number K_P which depends only on P. To determine K_P we choose an orthonormal basis $\{e_i\}$ for P. Then

$$f_{P;\bar{e}_i,e_j}(g) = M_{ij}(g),$$

where $M_{ij}(g)$ is the unitary matrix representing the action of g on P. Taking $\{\eta_1, \xi_1, \eta_2, \xi_2\} = \{\bar{e}_i, e_j, \bar{e}_k, e_l\}$ we find

$$\int_G \overline{M_{ij}(g)} M_{kl}(g) d(g) = K_P \delta_{ik} \delta_{jl}. \tag{9.17}$$

9 The Peter-Weyl theorem

But $\overline{M_{ij}}(g) = M_{ji}(g^{-1})$. Putting $i = k$ in (9.17), and summing over i, we get

$$\int_G M_{jl}(1)d(g) = K_P \dim(P)\delta_{jl}.$$

But $M_{jl}(1) = \delta_{jl}$, and $\int_G 1 d(g) = 1$, so $K_P = 1/\dim(P)$, as we want.

10
Functions on $\mathbb{R}^n$ and S^{n-1}

The spaces of functions on Euclidean space $\mathbb{R}^n$ and on the unit sphere S^{n-1} provide simple concrete illustrations of the representation theory we have been developing.

Let P_k denote the homogeneous polynomials of degree k on $\mathbb{R}^n$, with complex coefficients, and let F_k denote their restrictions to S^{n-1}. Thus

$$\mathbb{C}[x_1,\ldots,x_n] = \bigoplus_{k \geq 0} P_k,$$

while

$$P_k \cong F_k \supset F_{k-2} \supset F_{k-4} \supset \cdots,$$

because $\sum x_i^2 = \|x\|^2 = 1$ on S^{n-1}.

The space P_k is a representation of O_n, and we can introduce an invariant inner product. Let H_k be the orthogonal complement of F_{k-2} in F_k, so that

$$F_k = H_k \oplus F_{k-2} = H_k \oplus H_{k-2} \oplus H_{k-4} \oplus \cdots.$$

The space H_k is called the space of *spherical harmonics* of degree k on S^{n-1}. We shall see in a moment that it is an irreducible representation of O_n, and the reason for the name "harmonic" will also appear. If $n = 3$ it has dimension $2k + 1$, since in general $\dim(P_k) = \binom{n+k-1}{k}$.

The polynomials on $\mathbb{R}^n$ are dense in the space $C(S^{n-1})$ of continuous functions on S^{n-1} by the Stone-Weierstrass theorem. This implies

$$C(S^{n-1})^{\text{fin}} = \bigcup_{k \geq 0} F_k = \bigoplus_{k \geq 0} H_k. \tag{10.1}$$

Let us compare this decomposition with the assertion of Theorem (9.14), writing $C(S^{n-1})$ as $C(O_n/O_{n-1})$. If H_k is irreducible, then (10.1) and (9.14) are compatible if $H_k^{O_{n-1}}$ is one-dimensional for each k (for clearly we have $H_k = \overline{H}_k$ in this case). But, conversely, (9.14) implies that $\dim(\overline{P}^{O_{n-1}}) \geq 1$ for each irreducible representation P which occurs in $C(S^{n-1})$. So if we show that $\dim(H_k^{O_{n-1}}) = 1$ it will follow that H_k is irreducible.

Suppose that O_{n-1} is the subgroup of O_n which leaves the x_1-axis fixed. Then $f \in P_k$ belongs to $P_k^{O_{n-1}}$ only if it is a linear combination of the polynomials

$$x_1^k, x_1^{k-2}\rho, x_1^{k-4}\rho^2, \ldots,$$

where $\rho = x_2^2 + x_3^2 + \ldots + x_n^2$. So $\dim(P_k^{O_{n-1}}) = [k/2]$, and hence $\dim(H_k^{O_{n-1}}) = 1$. This proves that H_k is irreducible.

If $\varphi \in H_k$ then $r^k \varphi$ is a homogeneous polynomial of degree k on $\mathbb{R}^n$, where $r = \|x\|$. Let $\hat{H}_k$ denote the polynomials of this form, so that $\hat{H}_k \xrightarrow{\cong} H_k$ under restriction to S^{n-1}. We have proved that

$$P_k = \hat{H}_k \oplus r^2 \hat{H}_{k-2} \oplus r^4 \hat{H}_{k-4} \oplus \ldots,$$

or, alternatively, that any polynomial f on $\mathbb{R}^n$ can be decomposed

$$f(r\sigma) = \sum r^k f_k(r^2) \varphi_k(\sigma),$$

where $\sigma \in S^{n-1}$, $\varphi_k \in H_k$ and f_k is a polynomial in r^2. In other words

$$\mathbb{C}[x_1,\ldots,x_n] = \bigoplus_{k \geq 0} E_k \otimes H_k \qquad (10.2)$$

where E_k is the space of radial functions of the form $r^k f_k(r^2)$.

The *Laplacian* $\Delta = \sum(\frac{\partial}{\partial x_i})^2$ is a map

$$\Delta : \mathbb{C}[x_1,\ldots,x_n] \to \mathbb{C}[x_1,\ldots,x_n].$$

It commutes with the action of O_n, and maps P_k into P_{k-2}. By Schur's lemma it must map the irreducible subspace $\hat{H}_k$ of P_k to zero, as $\hat{H}_k$ does not occur in P_{k-2}. In fact

Proposition 10.3 *The space $\hat{H}_k$ is exactly the kernel of Δ restricted to P_k, i.e. it is the space of* harmonic polynomials *of degree k on $\mathbb{R}^n$.*

Proof. It is easy to check that

$$\Delta(r^2 \varphi) = 2(n+2k)\varphi + r^2 \Delta \varphi \qquad (10.4)$$

for any $\varphi \in P_k$, and hence, inductively, that

$$\Delta(r^{2s}\varphi) = 2s(n + 2k - 2s - 2)r^{2s-2}\varphi + r^{2s}\Delta\varphi$$

for $\varphi \in P_{k-2s}$. So Δ maps $r^{2s}\hat{H}_{k-2s}$ isomorphically to $r^{2s-2}\hat{H}_{k-2s}$ if $s > 0$.

It is well-known that the Laplacian Δ on $\mathbb{R}^n$ can be written

$$\Delta = (\frac{\partial}{\partial r})^2 + \frac{n-1}{r}\frac{\partial}{\partial r} + \frac{1}{r^2}\Delta_S, \tag{10.5}$$

where Δ_S is the Laplacian on S^{n-1}. As Δ_S commutes with the action of O_n it must preserve each subspace H_k, and must act on H_k by multiplication by a scalar. Because $\Delta(r^k\varphi) = 0$ when $\varphi \in H_k$ we can substitute in (10.5) to obtain

Proposition 10.6 *(i) The Laplacian Δ_S acts as $-k(k+n-2)$ on H_k.*
(ii) In terms of the decomposition (10.2) we have

$$\Delta = \sum_{k \geq 0} \Delta_k \otimes 1,$$

where $\Delta_k : E_k \to E_k$ is given by

$$\Delta_k = (\frac{\partial}{\partial r})^2 + \frac{n-1}{r}\frac{\partial}{\partial r} - \frac{k(k+n-2)}{r^2}.$$

There is a striking curiosity to be noticed here. On the polynomial ring $\mathbb{C}[x_1,\ldots,x_n]$ we have three operators

$$\frac{1}{2}\Delta, r\frac{\partial}{\partial r} + \frac{n}{2}, \frac{1}{2}r^2, \tag{10.7}$$

where $\frac{1}{2}r^2$ denotes the multiplication operator by $\frac{1}{2}r^2$. They all commute with the O_n-action, and can be regarded as operators on the spaces E_k. If one calls them $\{e, h, f\}$ one can rapidly check, using (10.4), that they satisfy

$$[h, e] = -2e, \quad [h, f] = 2f, \quad [e, f] = h,$$

i.e. they define a representation of the Lie algebra $sl_2\mathbb{R}$ of $SL_2\mathbb{R}$ on $\mathbb{C}[x_1,\ldots,x_n]$, with the harmonic polynomials as lowest-weight vectors. Each E_k is an irreducible representation of $sl_2\mathbb{R}$, as it is generated by the lowest-weight vector r^k, of weight $k + \frac{n}{2}$. This Lie algebra action does not come from an action of the group $SL_2\mathbb{R}$ on $\mathbb{C}[x_1,\ldots,x_n]$, and there is no prima facie geometrical reason for $sl_2\mathbb{R}$ to appear. I shall not pursue this any further, but I shall give a fuller account of a closely related situation in Chapter 17.

The Radon transform

An amusing application of the decomposition $C(S^{n-1}) = \hat{\bigoplus} H_k$ is to the *Radon transform*. This is the linear map

$$\mathcal{R} : C(S^{n-1}) \to C(S^{n-1})$$

defined by

$(\mathcal{R}f)(x) = $ (average of f over the great $(n-2)$-sphere with pole x).

As $\mathcal{R}f$ is automatically an even function on S^{n-1}, i.e. $\mathcal{R}f(-x) = \mathcal{R}f(x)$, and $\mathcal{R}f = 0$ if f is odd, it is best to think of $\mathcal{R}$ as a map

$$\mathcal{R} : C_+(S^{n-1}) \to C_+(S^{n-1}),$$

where C_+ denotes the even functions. We have $C_+(S^{n-1}) = \hat{\bigoplus} H_{2k}$.

One would like to know whether $\mathcal{R}$ is bijective, i.e. whether f can be reconstituted from its averages. Because $\mathcal{R}$ commutes with O_n it must map each H_{2k} to itself by multiplication by some scalar λ_{nk}. To calculate λ_{nk} we observe that it is the value of $\mathcal{R}f$ at $(1,0,\ldots,0)$ when $f = (x_1 + ix_2)^{2k} \in H_{2k}$. By explicit integration we get

$$\lambda_{nk} = \frac{(-1)^k}{\sqrt{\pi}} \frac{\Gamma(\frac{n-1}{2})\Gamma(k+\frac{1}{2})}{\Gamma(k+\frac{n-1}{2})}.$$

If $n = 3$ this is $\frac{1}{2^{2k}}\binom{2k}{k}$. In any case, it is non-zero, and decays like $1/k^{\frac{n}{2}-1}$ as $k \to \infty$. We conclude that the Radon transform is injective, and, by (9.13), bijective on C^∞ functions.

11
Induced representations

The Peter-Weyl theorem describes the G-action on $C(X)$ when $X = G/H$ is a homogeneous space of a compact group G. It works equally well if we want to study, say, the space $\text{Vect}(X)$ of tangent vector fields on X.

Let $T = T_{x_0}X$ be the tangent space to X at its base-point $x_0 = [H]$. The subgroup H acts on T, for each $h \in H$ defines a map $X \to X$ which leaves x_0 fixed and therefore induces $h : T \to T$.

Proposition 11.1 *Tangent vector fields ξ on G/H can be identified with maps $\hat{\xi} : G \to T$ which are H-equivariant in the sense that*

$$\hat{\xi}(gh^{-1}) = h\hat{\xi}(g).$$

The G-action on $\text{Vect}(G/H)$ corresponds to the action

$$(g.\hat{\xi})(x) = \hat{\xi}(g^{-1}x).$$

In symbols

$$\text{Vect}(G/H) \cong \text{Map}_H(G; T)$$

as representations of G.

Proof. The tangent vector $\xi(gH)$ lies in $T_{gH}(G/H)$, so $\hat{\xi}(g) = g^{-1}\xi(gH)$ belongs to T. It depends, however, on g and not just on the coset gH: in fact $\hat{\xi}(gh^{-1}) = h\hat{\xi}(g)$. Conversely, given $\hat{\xi} : G \to T$ we can define $\xi(gH) = g\hat{\xi}(g) \in T_{gH}$.

Now we can apply the Peter-Weyl theorem. As we have not discussed real representations we had better consider the complexification

$$\text{Vect}(G/H) \otimes \mathbb{C} \cong \text{Map}_H(G; T_{\mathbb{C}}),$$

11 Induced representations

where $T_{\mathbb{C}} = T \otimes \mathbb{C}$. We have

$$\text{Map}_H(G; T_{\mathbb{C}}) = \{C(G) \otimes T_{\mathbb{C}}\}^H$$
$$= \widehat{\bigoplus_P} P \otimes (\overline{P} \otimes T_{\mathbb{C}})^H,$$

where P runs through the irreducible representations of G. Because each P is unitary we have $\overline{P} \cong P^*$, and so $(\overline{P} \otimes T_{\mathbb{C}})^H$ can be identified with the space $\text{Hom}_H(P; T_{\mathbb{C}})$ of H-equivariant linear maps $P \to T_{\mathbb{C}}$.

Example. If $X = S^2 = O_3/O_2$ then $T \cong \mathbb{R}^2$ with the obvious action of O_2. The irreducible representations of O_3 are the spaces H_k of spherical harmonics, and $(H_k \otimes \mathbb{C}^2)^{O_2}$ is zero if $k = 0$, and is 2-dimensional if $k > 0$. So

$$\widehat{\text{Vect}(S^2) \otimes \mathbb{C}} \cong \widehat{\bigoplus_{k>0}}(H_k \oplus H_k).$$

There is no O_3-invariant vector field on S^2, so H_0 does not occur.

Representations of G of the form $\text{Map}_H(G; M)$, where M is a representation of H, are called *induced representations*; more precisely, $\text{Map}_H(G; M)$ is called the representation of G induced by the representation H of M. They are natural generalizations of the spaces $C(G/H)$ — i.e. the case $M = \mathbb{C}$ — and they can always be interpreted as the spaces of sections of G-*vector-bundles* on G/H analogous to the tangent bundle. A G-vector-bundle on a space X on which G acts is a family of vector spaces $\{E_x\}_{x \in X}$ together with linear maps $g : E_x \to E_{gx}$ for each $g \in G$ and $x \in X$. The family is required to be locally trivial in a sense I shall not discuss.

Example. Let X be complex projective space $\mathbb{P}_{\mathbb{C}}^{n-1} = \mathbb{P}(\mathbb{C}^n)$. This is a homogeneous space of U_n, and also one of $G = GL_n\mathbb{C}$. Let us think of it as G/H, where $H = GL_{1,n-1}(\mathbb{C})$ is the group of echelon matrices

$$\begin{pmatrix} h_{11} & h_{12} & \cdots & h_{1n} \\ 0 & h_{22} & \cdots & h_{2n} \\ 0 & h_{32} & \cdots & h_{3n} \\ \cdots & \cdots & \cdots & \cdots \\ 0 & h_{n2} & \cdots & h_{nn} \end{pmatrix}$$

Let V_k denote the space $(S^k \mathbb{C}^n)^*$ of homogeneous polynomials on $\mathbb{C}^n$ of degree k. A polynomial $p \in V_k$ is not a function on $\mathbb{P}_{\mathbb{C}}^{n-1}$: it assigns to each point L of $\mathbb{P}_{\mathbb{C}}^{n-1}$ (where L is a line in $\mathbb{C}^n$) a point $p(L)$ of the

1-dimensional space E_L of homogeneous functions of degree k on L, i.e. it is a section of the 1-dimensional complex vector bundle $E = \{E_L\}_{L \in \mathbb{P}_\mathbb{C}^{n-1}}$ on $\mathbb{P}_\mathbb{C}^{n-1}$. At first sight this may not seem a very fruitful way of regarding the polynomial p, but in the long term it is the right point of view. (See Chapter 14.)

Alternatively, p can be regarded as a map $\tilde{p} : G \to \mathbb{C}$ which satisfies

$$\tilde{p}(gh) = h_{11}^k \tilde{p}(g) \tag{11.2}$$

when $h \in H$ is as above. For applying (11.2) when h is of the form

$$\begin{pmatrix} 1 & * \\ 0 & * \end{pmatrix}$$

tells us that $\tilde{p}(g)$ depends only on the first column of g, i.e. that $\tilde{p}$ is a function $\mathbb{C}^n - \{0\} \to \mathbb{C}$; and then applying it when

$$h = \begin{pmatrix} \lambda & 0 \\ 0 & 1 \end{pmatrix}$$

tells us that $\tilde{p}$ is homogeneous of degree k.

In fact V_k consists precisely of all *holomorphic* functions $\tilde{p} : G \to \mathbb{C}$ satisfying (11.2). Whenever G and H are complex Lie groups it makes sense to speak of the representation $\text{Map}_H^{\text{hol}}(G; M)$ of G *holomorphically induced* from a holomorphic representation M of H (or, equivalently, of the holomorphic sections of the vector bundle E on G/H). So we can state what we have proved as

Proposition 11.3 *The representation $(S^k \mathbb{C}^n)^*$ of $GL_n\mathbb{C}$ is holomorphically induced from the 1-dimensional representation $h \mapsto h_{11}^{-k}$ of $GL_{1,n-1}\mathbb{C}$.*

For future use let me point out that if $\tilde{p} : GL_n\mathbb{C} \to \mathbb{C}$ is a holomorphic function then to prove that $\tilde{p}(g)$ is a homogeneous polynomial function of the first column of g we need to assume (11.2) only for h in the subgroup B of upper-triangular matrices. One can prove this by an elementary explicit argument, but the point is that the homogeneous space $GL_{1,n-1}/B$ is a *compact* complex manifold, and so any holomorphic function on it is constant. In any case, beside (11.3) we have

Proposition 11.4 *The representation $(S^k \mathbb{C}^n)^*$ of $GL_n\mathbb{C}$ is holomorphically induced from the 1-dimensional representation $h \mapsto h_{11}^{-k}$ of B.*

11 Induced representations

We shall prove in Chapter 14 that *every* holomorphic representation of $GL_n\mathbb{C}$ is holomorphically induced from a 1-dimensional representation

$$h \longmapsto h_{11}^{k_1} \ h_{22}^{k_2} \ldots h_{nn}^{k_n}$$

of B for some $\mathbf{k} = (k_1, \ldots, k_n) \in \mathbb{Z}^n$.

12
The complexification of a compact group

The unitary group U_n is a maximal compact subgroup of the complex Lie group $GL_n\mathbb{C}$. There are three other important aspects of the relationship between these two groups.

The first is very obvious. The Lie algebra of U_n is the space of $n \times n$ skew hermitian matrices, and any matrix can be expressed uniquely in the form $A + iB$ with A and B skew hermitian. This gives us

Proposition 12.1 *The Lie algebra $M_n\mathbb{C}$ of $GL_n\mathbb{C}$ is the complexification of the Lie algebra of U_n.*

This is what is usually meant by saying that the group $GL_n\mathbb{C}$ is the *complexification* of the group U_n.

Proposition 12.2 *The algebra $C_{\mathrm{alg}}(U_n)$ of representative functions on U_n is precisely the algebra $\mathbb{C}[a_{ij}, \Delta^{-1}]$ of polynomial functions on the algebraic group $GL_n\mathbb{C}$, where $\Delta = \det(a_{ij})$.*

Proof. We have already explained that, by the Stone-Weierstrass theorem, the representative functions on U_n are the polynomials in a_{ij} and $\overline{a_{ij}}$. But, being unitary, $(\overline{a_{ij}})$ is the transposed inverse matrix to (a_{ij}), so $\overline{a_{ij}} = p_{ij}/\Delta$, where p_{ij} is a polynomial in the a_{ij}.

Finally, we have

Proposition 12.3 *Every representation of U_n is the restriction of a unique holomorphic representation of $GL_n\mathbb{C}$.*

Proof. The uniqueness is because a holomorphic map

$$GL_n\mathbb{C} \to \mathrm{Aut}(V)$$

12 The complexification of a compact group

is determined by its values on U_n (e.g. a holomorphic function on $\mathbb{C}-\{0\}$ is determined by its values on the unit circle $\mathbb{T}$). The extendability is because $C_{\text{alg}}(U_n) = \mathbb{C}[a_{ij}, \Delta^{-1}]$: a more detailed argument will be given in the proof of (13.2) in the next section.

Remarks.

(i) It is not true that any continuous function on U_n extends to a holomorphic function on $GL_n\mathbb{C}$, but the ones which do extend are dense in $C(U_n)$.

(ii) For the standard representation $V = \mathbb{C}^n$ of U_n the complex-conjugate representation $\overline{V}$ is isomorphic to the dual representation V^*. But $\overline{V}$ and V^* are *not* equivalent representations of $GL_n\mathbb{C}$: for V^* is holomorphic and $\overline{V}$ is not.

(iii) The holomorphic representations of $GL_n\mathbb{C}$ are a very different thing from its unitary representations, which are all infinite dimensional. A non-trivial holomorphic representation cannot be unitary, for if $f : X \to GL_n\mathbb{C}$ is holomorphic, where X is any complex manifold X, we cannot have $f(X) \subset U_N$. (Otherwise, using the local chart on $GL_n\mathbb{C}$ given by the logarithm, we should have a holomorphic map into $M_N\mathbb{C}$ with values contained in the real vector subspace of skew hermitian matrices.)

There is a complexification $G_{\mathbb{C}}$ of any compact group G, and all four of the above characterizations of the relationship between G and $G_{\mathbb{C}}$ continue to hold. In the language of Macdonald's lectures, $G_{\mathbb{C}}$ is the linear algebraic group whose coordinate ring is $C_{\text{alg}}(G)$.

Example. The complexification of O_n is

$$O_n(\mathbb{C}) = \{A \in GL_n\mathbb{C} : A^t A = 1\}.$$

13
The unitary groups and the symmetric groups

Weyl's correspondence

Hermann Weyl showed that the irreducible representations of the unitary group U_n are realized as spaces of "tensors" with various symmetry properties. This is a very beautiful and important theorem, with many ramifications. It is one of the starting-points of the modern theory of "quantum groups".

The most obvious representation of U_n is its natural action on $\mathbb{C}^n = V$. This induces an action on

$$V^{\otimes k} = V \otimes \cdots \otimes V$$

for each k. An element of $V^{\otimes k}$ is a "tensor", i.e. an array of numbers $\mathbf{a} = \{a_{i_1 i_2 \ldots i_k}\}$, with $1 \leq i_r \leq n$. A matrix $(u_{ij}) \in U_n$ acts on $V^{\otimes k}$ by $\mathbf{a} \mapsto \tilde{\mathbf{a}}$, where

$$\tilde{a}_{i_1 \ldots i_k} = \sum_{j_1, \ldots, j_k} u_{i_1 j_1} u_{i_2 j_2} \cdots u_{i_k j_k} a_{j_1 \ldots j_k} .$$

The representations $V^{\otimes k}$ of U_n are reducible. An element of $V \otimes V$ can be written as the sum of a symmetric and a skew tensor

$$V \otimes V = S^2 V \oplus \Lambda^2 V ,$$

and both $S^2 V$ and $\Lambda^2 V$ are irreducible representations of U_n. The case of $V^{\otimes 3}$ is a little more complicated. We have

$$V \otimes V \otimes V = S^3 V \oplus \Lambda^3 V \oplus W ,$$

where W consists of the a_{ijk} such that

$$a_{ijk} + a_{jki} + a_{kij} = 0 .$$

Both S^3V and Λ^3V are irreducible under U_n, but W breaks into two irreducible representations $W = W_+ \oplus W_-$, where

$$W_\pm = \{(a_{ijk}) \in W : a_{ijk} = \pm a_{jik}\} \ .$$

There is a better way of expressing this. The space

$$Q = \{(\lambda, \mu, \nu) \in \mathbb{C}^3 : \lambda + \mu + \nu = 0\}$$

is an irreducible representation of the symmetric group S_3, and each of the representations $W_\pm$ is equivalent to the representation of U_n on the space

$$V_Q = \mathrm{Hom}_{S_3}(Q; V^{\otimes 3}) \ ,$$

of S_3-equivariant linear maps $Q \to V^{\otimes 3}$. (Here S_3 acts on $V^{\otimes 3}$ by permuting the factors. This action commutes with the action of U_n on $V^{\otimes 3}$, so U_n acts on V_Q.) We get isomorphisms $V_Q \to W_\pm$ by $f \mapsto f(q_\pm)$, where $q_+ = (1, 1, -2)$ and $q_- = (1, -1, 0)$. In fact the obvious map

$$Q \otimes V_Q \longrightarrow W$$

of representations of $S_3 \times U_n$ is an isomorphism. The decomposition $W = W_+ \oplus W_-$ corresponds to writing $Q = \mathbb{C} \oplus \mathbb{C}$ by using the basis $\{q_\pm\}$ for Q, which is singled out as the basis of eigenvectors for the transposition $(12) \in S_3$.

In general the symmetric group S_k acts on $V^{\otimes k}$, and we already know that we can decompose $V^{\otimes k}$ under S_k

$$V^{\otimes k} = \bigoplus_Q Q \otimes V_Q \ , \tag{13.1}$$

where Q runs through the irreducible representations of S_k and

$$V_Q = \mathrm{Hom}_{S_k}(Q; V^{\otimes k}) \ .$$

As before, U_n acts on V_Q, so (13.1) is an isomorphism of representations of $S_k \times U_n$. We call V_Q the "tensors of degree k with symmetry of type Q".

Weyl's theorem is

Theorem 13.2 *V_Q is an irreducible representation of U_n, and, up to multiplication by a power of the determinant, all irreducible representations of U_n*

arise in this way for some k. Furthermore, all irreducible representations of S_k occur in $V^{\otimes k}$ if $\dim(V) \geq k$.

The theorem establishes a 1-1 correspondence between the irreducible representations of U_n contained in $V^{\otimes k}$ and the irreducible representations of S_k contained in $V^{\otimes k}$. We shall see in the next section that irreducible representations of U_n, in general, are indexed by sequences of integers $k_1 \geq k_2 \geq \cdots \geq k_n$. The corresponding representation is contained in $V^{\otimes k}$ if $\sum k_i = k$ and $k_n \geq 0$. Multiplying by the determinant changes $(k_1, \ldots, k_n)$ to $(k_1 + 1, \ldots, k_n + 1)$.

Proof of (13.2). Essentially we have already seen why all representations of U_n arise. For the matrix entries of all subrepresentations of all $V^{\otimes k}$ form a subalgebra A of $C_{\text{alg}}(U_n)$ which would be orthogonal to the matrix entries of any hypothetical "missing" representation. But (if we put in arbitrary powers of the determinant Δ) the algebra A certainly contains $\mathbb{C}[a_{ij}, \Delta^{-1}]$, which is dense in $C(U_n)$.

It is also clear why all representations of S_k occur in $V^{\otimes k}$ if $k \geq n$. For they all occur in the left-action of S_k on the group-ring $\mathbb{C}[S_k]$, and if $\{e_i\}$ is a basis of V the orbit of $e_{i_1} \otimes \cdots \otimes e_{i_k} \in V^{\otimes k}$ under S_k spans a copy of $\mathbb{C}[S_k]$ if all the e_i occur among the e_{i_r}.

The hardest part is to prove that V_Q is irreducible. The argument is short, but extremely ingenious. We know from Chapter 12 that it is enough to show that V_Q is indecomposable under the action of $G = GL_n\mathbb{C}$, and hence to show that $\text{End}_G(V_Q) = \mathbb{C}$. Now $\mathbb{C}$ is the *centre* of the matrix algebra $\text{End}(V_Q)$, and by Schur's lemma we have

$$\text{End}_{S_k}(V^{\otimes k}) = \bigoplus_Q \text{End}(V_Q)$$

and

$$\text{End}_{G \times S_k}(V^{\otimes k}) = \bigoplus_Q \text{End}_G(V_Q)$$

It is therefore enough to show that the algebra $\text{End}_{G \times S_k}(V^{\otimes k})$ is contained in the centre of $\text{End}_{S_k}(V^{\otimes k})$, and hence to show that the image of G in $\text{End}_{S_k}(V^{\otimes k})$ spans it as a vector space.

But

$$\text{End}_{S_k}(V^{\otimes k}) = (\text{End}(V)^{\otimes k})^{S_k},$$

where $(\text{End}(V)^{\otimes k})^{S_k}$ denotes the part invariant under S_k. A linear form

13 The unitary and symmetric groups

on this is simply† a homogeneous polynomial of degree k on the vector space $\text{End}(V)$. Such a polynomial certainly vanishes if it vanishes on the dense open subset G of $\text{End}(V)$. That completes the proof.

Quantum groups

In recent years the relation between the representations of the unitary and symmetric groups has become the height of fashion, because of the emerging theory of "quantum groups". I cannot describe this development here. It must suffice to say that the action of the Lie algebra $\mathfrak{u}_n$ of U_n on $(\mathbb{C}^n)^{\otimes k}$, and the commuting action of S_k on it, each possess a canonical 1-parameter family of deformations, indexed by $q \in \mathbb{C}^\times$. The deformation is not, however, through representations of $\mathfrak{u}_n$ and S_k. The action of S_k is deformed to a representation of the *braid group* Br_k on k strings. A *braid* is an "enhanced" permutation, described by a diagram like

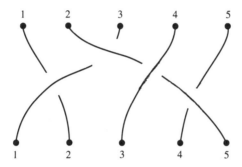

For each $q \in \mathbb{C}^\times$ there is an action of the group Br_k on $(\mathbb{C}^n)^{\otimes k}$, and only when $q = 1$ does the action factorize through S_k. Meanwhile, the deformed action of $\mathfrak{u}_n$ is an action of a new kind of algebraic object $(\mathfrak{u}_n)_q$ which depends on q. This is the so-called *quantum group*. It remains true after the deformation that the actions of $(\mathfrak{u}_n)_q$ and of Br_k are commutants of each other, and so either can be used to construct the other.

The representations of the braid groups are of great interest in three-

---

† We are using the fact that for any representation W of S_k a linear form on W^{S_k} comes from an S_k-invariant linear form on W. This holds because a linear form on W^{S_k} can be extended arbitrarily to a linear form on W, and then averaged over S_k. We are also using the fact that a homogeneous polynomial of degree k on a vector space E is the same as a symmetric k-fold multilinear map $E \times \cdots \times E \to \mathbb{C}$, i.e. an S_k-invariant linear form on $E^{\otimes k}$.

dimensional topology, while those of the quantum group arise in two-dimensional quantum field theory and statistical mechanics. The link so established between these areas has been one of the great surprises of recent mathematics.

14
The Borel-Weil theorem

I shall now give a systematic description of all the irreducible unitary representations V of U_n, in a form which generalizes immediately to any compact group.

We begin by observing that the U_n-action on V extends to a holomorphic action of $G_{\mathbb{C}} = GL_n\mathbb{C}$.

Then we decompose V under the action of the subgroup $T = \mathbb{T}^n$ of diagonal matrices

$$u = \begin{pmatrix} u_1 & & \\ & \ddots & \\ & & u_n \end{pmatrix}.$$

Because T is commutative and acts unitarily we can find a basis of V consisting of *weight vectors*, i.e. vectors v which are eigenvectors of the T-action. For such a vector we have $uv = \lambda(u)v$, where

$$\lambda(u) = u_1^{k_1} u_2^{k_2} \cdots u_n^{k_n}$$

for some multi-index $\mathbf{k} = (k_1, \cdots, k_n) \in \mathbb{Z}^n$ called the *weight* of v.

The Lie algebra $\mathfrak{g}_{\mathbb{C}}$ of $GL_n\mathbb{C}$ acts on V. Let $E_{ij} \in \mathfrak{g}_{\mathbb{C}} = M_n\mathbb{C}$ be the matrix with 1 in the (i,j) place and 0 elsewhere. If v is a weight vector with weight k then, because $uE_{ij}u^{-1} = u_i u_j^{-1} E_{ij}$, the vector $E_{ij}v$ is either a weight vector or zero, and its weight is $\mathbf{k} + \epsilon_{ij}$, where $\epsilon_{ij} \in \mathbb{Z}^n$ has 1 (resp. -1) in the ith (resp. j^{th}) place.

By ordering the weights lexicographically we can find a *highest weight vector* $v \in V$, which has the property that $E_{ij}v = 0$ where $i < j$. It will turn out that there is only one such vector, up to a scalar multiple, but we do not know that yet.

Let B be the subgroup of upper triangular matrices in $G_{\mathbb{C}}$: its Lie algebra is spanned by the E_{ij} with $i < j$ together with the diagonal matrices. So the highest weight vector v is an eigenvector of B, i.e.

$$bv = \lambda(b)v,$$

where $\lambda : B \to \mathbb{C}^{\times}$ is the homomorphism such that $\lambda(b) = b_{11}^{k_1} \ldots b_{nn}^{k_n}$ if

$$b = \begin{pmatrix} b_{11} & & * \\ & \ddots & \\ 0 & & b_{nn} \end{pmatrix}.$$

Now define a $G_{\mathbb{C}}$-map

$$V^* \to M_{-\mathbf{k}} = \operatorname{Map}_B^{\mathrm{hol}}(G_{\mathbb{C}} : \mathbb{C})$$

by

$$\alpha \mapsto \{g \mapsto \alpha(gv)\},$$

where the right-hand side means the holomorphic maps $f : G_{\mathbb{C}} \to \mathbb{C}$ such that $f(gb) = \lambda(b)f(g)$. In other words, $M_{-\mathbf{k}}$ is the representation of $G_{\mathbb{C}}$ *holomorphically induced* from the 1-dimensional representation λ^{-1} of B. The Borel-Weil theorem is

Theorem 14.1 *The map* $V^* \to M_{-\mathbf{k}}$ *is an isomorphism, and so*

$$V \cong M^*_{-\mathbf{k}}.$$

Proof. The map is clearly non-zero, and hence injective, as V^* is irreducible. To prove it is surjective we show $M_{-\mathbf{k}}$ is also irreducible. It is enough to see that $M_{-\mathbf{k}}$ contains at most *one* vector fixed under the action of the subgroup $\tilde{N}$ of strictly lower-triangular matrices, for if it split into two pieces each would contain a lowest weight vector, and such a vector is fixed under $\tilde{N}$.

If $f \in M_{-\mathbf{k}}$ is fixed under $\tilde{N}$ we have $f(\tilde{n}b) = \lambda(b)f(1)$. So $f|\tilde{N}B$ is completely determined by its value $f(1) \in \mathbb{C}$. But an open dense subset of elements $g \in G_{\mathbb{C}}$ can be factorized $g = \tilde{n}b$ with $\tilde{n} \in \tilde{N}$ and $b \in B$ (I shall discuss this further below), so f itself is completely determined by $f(1)$, and $\dim(M_{-\mathbf{k}}^{\tilde{N}}) \leq 1$, as we want.

A corollary of the proof of the Borel-Weil theorm is

Theorem 14.2 *Any irreducible representation of* U_n *contains a unique highest weight vector.*

14 The Borel-Weil theorem

The Borel-Weil theorem constructs and labels all the irreducible representations of U_n. They are in $1-1$ correspondence with their highest weights $\mathbf{k} = (k_1, \ldots, k_n) \in \mathbb{Z}^n$. A possible highest weight must be *dominant*, i.e.

$$k_1 \geqslant k_2 \geqslant \cdots \geqslant k_n,$$

for if $\mathbf{k}$ is a weight of a representation V then so is any $\mathbf{k}'$ got by reordering the k_i. We shall see below that if $\mathbf{k}$ is not dominant then the holomorphically induced representation $M_{-\mathbf{k}}$ is zero. We should also mention that the dual representation $M^*_{-\mathbf{k}}$ is easily seen to be isomorphic to $M_\mathbf{l}$, where

$$\mathbf{l} = (k_n, k_{n-1}, \ldots, k_1),$$

for $M_\mathbf{l}$ is an irreducible representation whose lowest weight is

$$\mathbf{l} = (l_1, \ldots, l_n),$$

and whose highest weight is therefore necessarily $(l_n, l_{n-1}, \ldots, l_1)$.

For any representation V of $G_\mathbb{C}$ the orbit of the highest weight ray $[v]$ in the projective space $\mathbb{P}(V)$ under the action of $G_\mathbb{C}$ gives a holomorphic map

$$i : G_\mathbb{C}/B \to \mathbb{P}(V).$$

For most representations V the subgroup B is precisely the stabilizer of the ray $[v]$, and so i is an embedding. But if $\lambda : B \to \mathbb{C}^\times$ extends to a larger subgroup P containing B then P stabilizes $[v]$, and the map factorizes $G_\mathbb{C}/B \to G_\mathbb{C}/P \to \mathbb{P}(V)$. In this case the representation V^* is holomorphically induced from P, as well as from B. We saw in Chapter 11 that this occurs when $V = (S^k \mathbb{C}^n)^*$ and $P = GL_{1,n-1}\mathbb{C}$. Subgroups of $G_\mathbb{C}$ containing B are called *parabolic subgroups*.

Example. The representation $(\bigwedge^k \mathbb{C}^n)^*$ of $GL_n\mathbb{C}$ is induced from the representation

$$\begin{pmatrix} a & b \\ 0 & d \end{pmatrix} \mapsto \det(a)^{-1}$$

of $P = GL_{k,n-k}$, or alternatively from $\lambda : B \to \mathbb{C}^\times$, where

$$\lambda(b) = (b_{11}b_{22}\cdots b_{kk})^{-1}.$$

For if $f : G_\mathbb{C} \to \mathbb{C}$ satisfies $f(gp) = f(g)$ when

$$p = \begin{pmatrix} 1 & * \\ 0 & * \end{pmatrix} \in P$$

then $f(g)$ depends only on the first k columns $g_1, \cdots, g_k$ of g. If also $f(gp) = \det(a)^{-1} f(g)$ when

$$p = \begin{pmatrix} a & 0 \\ 0 & 1 \end{pmatrix}$$

then $f(g)$ depends only on $g_1 \wedge g_2 \wedge \cdots \wedge g_k$. So f is a holomorphic section of the line bundle on the Grassmannian $Gr_k(\mathbb{C}^n) = G_\mathbb{C}/P$ whose fibre at a subspace W of $\mathbb{C}^n$ is $(\bigwedge^k W)^*$. On the other hand, an element of $(\bigwedge^k V)^*$ gives an element of $(\bigwedge^k W)^*$ for each $W \in Gr_k(\mathbb{C}^n)$ by restriction.

The embedding

$$Gr_k(\mathbb{C}^n) \to \mathbb{P}(\bigwedge^k \mathbb{C}^n)$$

given by $W \mapsto [w_1 \wedge \cdots \wedge w_k]$, where $\{w_i\}$ is a basis for W, is called the *Plücker embedding*, and the $\binom{n}{k}$ components of $w_1 \wedge \cdots \wedge w_k$ are called the Plücker coordinates of W.

It is worth analysing the general holomorphically induced representation $M_{-\mathbf{k}}$ of $G_\mathbb{C} = GL_n\mathbb{C}$ a little more explicitly, in order to relate it to Weyl's tensorial construction described in Chapter 13. By multiplying by a power of the determinant we can reduce to the case where $k_n = 0$. Then we have

Theorem 14.3 *(i) The elements $f : G_\mathbb{C} \to \mathbb{C}$ of $M_{-\mathbf{k}}$, when regarded as functions $f(\xi_1, \ldots, \xi_n)$ of n vectors, depend polynomially on each vector ξ_i, and are homogeneous of degree k_i in ξ_i. In particular, $M_{-\mathbf{k}}$ is a subrepresentation of*

$$(S^{k_1}\mathbb{C}^n)^* \otimes \cdots \otimes (S^{k_{n-1}}\mathbb{C}^n)^*.$$

(ii) $M_{-\mathbf{k}} = 0$ unless $\mathbf{k}$ is dominant, i.e. $k_1 \geqslant \cdots \geqslant k_n$.

Proof. (i) If $k_n = 0$ then an argument we have already used in Chapter 11 shows that $f(\xi_1, \ldots, \xi_n)$ is independent of ξ_n. Now hold all ξ_j fixed for $j \neq i$, and consider the dependence on ξ_i. We get a holomorphic function $f_{(i)}$ defined on $\mathbb{C}^n - W$, where W is the $(n-2)$-dimensional subspace of $\mathbb{C}^n$ spanned by the ξ_j for $j \neq i, n$. By Hartogs's theorem ([Griffiths and Harris] page 7), which tells us that a holomorphic function cannot have singularities of codimension > 1, we conclude that $f_{(i)}$ extends to all of $\mathbb{C}^n$. As it is homogeneous, it must be a polynomial of degree k_i.

(ii) Restrict f to the copy of $GL_2\mathbb{C}$ in $GL_n\mathbb{C}$ formed by the ith and j^{th}

14 The Borel-Weil theorem

rows and columns. We get a function of a 2×2 matrix

$$\begin{pmatrix} a & b \\ c & d \end{pmatrix}$$

which is homogeneous of degrees k_i and k_j in the columns, and depends on the second column only through the determinant $ad - bc$. It is therefore of the form

$$p(a,c)(ad - bc)^{k_j},$$

where p is a homogeneous polynomial of degree $k_i - k_j$. In particular, $k_i \geqslant k_j$.

Everything said in this section generalizes from U_n to any compact group G. The diagonal matrices T are replaced by a maximal torus T, and the upper-triangular matrices B by the group B generated by $T_\mathbb{C}$ and the 1-parameter subgroups corresponding to the positive root-vectors in the Lie algebra of $G_\mathbb{C}$. The homogeneous space $G_\mathbb{C}/B$ is always a compact complex algebraic variety, and, as I mentioned in Chapter 4C, it possesses a beautiful decomposition into cells, called the *Bruhat decomposition*. The interplay between the geometry of these cells and the structure of the representations of $G_\mathbb{C}$ is fundamental in representation theory.

I shall give one more example of the Borel-Weil construction.

Example. The **spin representation** of O_{2n}.

We saw on page 61 that the homogeneous space $\mathfrak{I}_n = O_{2n}/U_n$ of complex structures on $\mathbb{R}^{2n}$ has a complex description as $O_{2n}(\mathbb{C})/P$, where P is a parabolic subgroup.

The group O_{2n} has a double covering $\tilde{O}_{2n}$, whose restriction to the subgroup U_n is the group $\tilde{U}_n$ of pairs (u, λ) with $u \in U_n$ and $\lambda^2 = \det(u) \in \mathbb{C}^\times$. Using the complex structure of $O_{2n}/U_n = \tilde{O}_{2n}/\tilde{U}_n$ we can form the representation of $\tilde{O}_{2n}$ holomorphically induced from the one-dimensional representation L^* of $\tilde{U}_n$ given by $(u, \lambda) \mapsto \lambda^{-1}$. This is the spin representation. It has dimension 2^n, and when restricted to $\tilde{U}_n$ it becomes

$$(\bigwedge \mathbb{C}^n) \otimes L^*.$$

More details of the construction of the spin representation from this point of view can be found in Chapter 12 of [Pressley and Segal]. I shall give a more conventional description of it in Chapter 17.

15
Representations of non-compact groups

This is a huge subject, and I can only make some orientational remarks. For excellent comprehensive introductions to the subject, from contrasting standpoints, I recommend the books by Kirillov and Knapp.

The first problem one meets is that infinite dimensional representations of a group come in families of roughly equivalent representations which one wants to lump together. For example, $PSU_{1,1}$ acts on the circle S^1, and hence on the functions on S^1. But one might want to consider its action on the continuous functions $C(S^1)$, the smooth functions $C^\infty(S^1)$, the L^2-functions $L^2(S^1)$, or perhaps some other class. For most purposes these representations of $PSU_{1,1}$ are not interestingly different. All of them are irreducible modulo the subspace $\mathbb{C}$ of constant functions, in the sense that when they are given their natural topologies there is no *closed* invariant subspace other than $\mathbb{C}$. If a representation can be made unitary it is natural to consider the Hilbert space version (which is unique if the representation is irreducible), but that has the disadvantage that the Lie algebra of the group does not act on it. For example, the 1-parameter group of rigid rotations of S^1 is generated by the Lie algebra element which acts on $C^\infty(S^1)$ by the differentiation operator $\frac{d}{d\theta}$. But $\frac{d}{d\theta}$ is not everywhere defined on $C(S^1)$ or $L^2(S^1)$.

There is a standard way to deal with this problem in the case of semisimple groups G. When given a representation of G on a topological vector space V we first consider the action on V of the maximal compact subgroup K of G. As we saw in Chapter 9 this picks out the dense subspace V^{fin} of *K-finite* vectors. In $PSU_{1,1}$ the maximal compact subgroup is the group $\mathbb{T}$ of rigid rotations of S^1, and we get the same space V^{fin} whether we start with $V = C^\infty(S^1)$ or $V = C(S^1)$ or $V = L^2(S^1)$: in each case V^{fin} consists of the *trigonometric polynomials*, i.e. the Fourier series

15 Representations of non-compact groups

$\Sigma a_k e^{ik\theta}$ with only finitely many a_k non-zero. Usually the group G does not act on V^{fin}. In our example, a typical element of $PSU_{1,1}$ takes the function $e^{i\theta}$ to

$$\frac{ae^{i\theta}+b}{\overline{b}e^{i\theta}+\overline{a}},$$

which is not a trigonometric polynomial. On the other hand, if M is a finite dimensional K-invariant subspace of V then so is $\mathfrak{g}M$, where $\mathfrak{g}$ is the Lie algebra of G. So we have

Proposition 15.1 *The Lie algebra $\mathfrak{g}$ of G acts on V^{fin}.*

In our example, the basis elements of the Lie algebra act by $\frac{d}{d\theta}$ and $e^{\pm i\theta}\frac{d}{d\theta}$.

A less obvious result, which I shall not prove, is

Proposition 15.2 *If an irreducible representation V of a semisimple group G is decomposed into isotypical parts*

$$V^{\text{fin}} = \bigoplus_P V_P$$

under the action of the maximal compact subgroup K of G, then $\dim(V_P)$ is finite for each P, i.e. each irreducible representation P of K occurs with finite multiplicity.

For semisimple groups the spaces V^{fin} with their simultaneous action of K and $\mathfrak{g}$ seem to be the right objects to study and classify. Passing from V to V^{fin} eliminates most of the analysis from the picture, and reduces the representation theory to algebra. For groups which are not semisimple the position is quite different, and (15.2) is far from true, as we shall see when discussing the Heisenberg group in Chapter 17.

The idea of focussing on the $(K,\mathfrak{g})$-action on V^{fin} is a fundamental step for another reason too: we should get very little information by considering the $\mathfrak{g}$-action alone. The representation theory of a group G and of its Lie algebra $\mathfrak{g}$ are not at all well related when infinite dimensional representations are considered. For example, the Lie algebras of $SL_2\mathbb{R}$ and SU_2 have the same complexification, and hence the same representation theory, but the representations of the two groups are quite different from each other. Most infinite dimensional representations of the Lie algebra $\mathfrak{g}$ do not come from representations of the group G. The basic example arises when G acts on a smooth manifold X. The group

G acts on $C^\infty(X)$, but if Y is an open subset of X not stable under G it certainly does not act on $C^\infty(Y)$. Nevertheless the Lie algebra $\mathfrak{g}$ acts on $C^\infty(Y)$, because functions on Y can be differentiated along the vector fields which generate the G-action.

Examples (i) The additive group $\mathbb{R}$ acts on $C^\infty(\mathbb{R})$ by translation, and the Lie algebra generator acts by $\frac{d}{dx}$, which acts on $C^\infty(a,b)$ for any open interval $(a,b) \subset \mathbb{R}$. But the group $\mathbb{R}$ does not act on $C^\infty(a,b)$.

(ii) If $G_\mathbb{C}$ is a complex algebraic group the Borel-Weil theorem (see Chap. 14) constructs the irreducible representations as

$$\mathrm{Map}_B^{\mathrm{hol}}(G_\mathbb{C}; \mathbb{C}_\lambda).$$

In the group $G_\mathbb{C}$ there is an important dense open subset $U = \tilde{N}B$. The spaces

$$\mathrm{Map}_B^{\mathrm{hol}}(U; \mathbb{C}_\lambda)$$

are called *dual Verma modules*. They are representations of $\mathfrak{g}_\mathbb{C}$ but not of $G_\mathbb{C}$.

A central role in the representation theory of any group G is played by the *Plancherel theorem*, which describes the decomposition of the Hilbert space $L^2(G)$ under the left- and right-action of G. For compact groups this was accomplished by the Peter-Weyl theorem, which enabled us to write any $f \in L^2 G$ as a sum

$$f = \sum_P f_P \tag{15.3}$$

of functions transforming according to the unitary irreducible representations P of G. The main difference in the non-compact case is that the sum (15.3) must be replaced by an integral

$$f = \int f_P d\mu(P) \tag{15.4}$$

with respect to a measure $d\mu(P)$ on the space of irreducible representations, just as Fourier series are replaced by Fourier integrals when one passes from the compact group $\mathbb{T}$ to the non-compact group $\mathbb{R}$.

As for Fourier integrals the Plancherel theorem tells us that

$$\|f\|^2 = \int \|f_P\|^2 \, d\mu(P).$$

15 Representations of non-compact groups

This formula, which generalizes the result

$$\| f \|^2 = \sum \frac{1}{\dim(P)} \| f_P \|^2$$

for compact groups, accounts for the name of the theorem.

16
Representations of $SL_2\mathbb{R}$

The representation theory of $SL_2\mathbb{R}$ is, of course, simpler than that of an arbitrary semisimple group, but nevertheless it exhibits the main features of the general case. In this section I shall describe the most important irreducible unitary representations of $G = SL_2\mathbb{R}$, namely the ones which are needed for the Plancherel theorem, or, equivalently, which "occur" in $L^2(SL_2\mathbb{R})$. (It would be more accurate to say that the representations not described form a set of measure zero for the Plancherel measure $d\mu(P)$ of (15.4).)

A natural family of representations to consider are those induced from the subgroup B of upper-triangular matrices. Then G/B is the real projective line $\mathbb{P}^1_\mathbb{R} \cong S^1$. The 1-dimensional representations of B are of the form

$$\begin{pmatrix} a & b \\ 0 & a^{-1} \end{pmatrix} \longmapsto (\text{sign}(a))^\epsilon \mid a \mid^p,$$

where $p \in \mathbb{C}$ and $\epsilon = 0$ or 1. The induced representation of G will be denoted by $E_{p,\epsilon}$. If $\epsilon = 0$ its elements are "$\frac{p}{2}$-densities" $f(\theta) \mid d\theta \mid^{p/2}$ on the circle $S^1 = G/B$, for B acts on the tangent space to G/B at its base-point by

$$\begin{pmatrix} a & b \\ 0 & a^{-1} \end{pmatrix} \longmapsto a^2.$$

If $\epsilon = 1$ the elements of $E_{p,\epsilon}$ are "twisted $\frac{p}{2}$- densities" $f(\theta)|d\theta|^{p/2}$ on S^1, where f is not a function but a cross-section of the Möbius band.

As representations of the maximal compact subgroup $\mathbb{T} = SO_2$ the spaces $E_{p,\epsilon}$ are independent of p: for $G/B = \mathbb{T}/\{\pm 1\}$, and $E_{p,\epsilon}$ is simply the space of functions $\varphi : \mathbb{T} \to \mathbb{C}$ such that

$$\varphi(-z) = (-1)^\epsilon \varphi(z).$$

Each isotypical piece for the $\mathbb{T}$-action is 1-dimensional. If p is not an integer then $E_{p,\epsilon}$ is an irreducible representation of G, as is easily seen by considering the action of the Lie algebra $\mathfrak{g}$ on $(E_{p,\epsilon})^{\text{fin}}$. But it is not usually a unitary representation. There is an obvious multiplication map

$$\bar{E}_{p,\epsilon} \times E_{p,\epsilon} \longrightarrow E_{p+\bar{p},0}.$$

The space of densities on S^1 is $E_{2,0}$, so when $p + \bar{p} = 2$ there is an invariant inner product on $E_{p,\epsilon}$ given by

$$\langle f_1, f_2 \rangle = \int \bar{f}_1 f_2.$$

Taking $p = 1 + is$, and either value of ϵ, we get two families of irreducible unitary representations parametrized by the real number s. They are called the *principal series* representations. (Actually $E_{1,1}$ should be excluded, as it is reducible : we shall see in a moment that $E_{1,1} \cong \Omega_{\text{hol}}^{\frac{1}{2}} \oplus \bar{\Omega}_{\text{hol}}^{\frac{1}{2}}$.)

The other important unitary representations are the *discrete series*, which are holomorphically induced from the compact subgroup $\mathbb{T} = SO_2$. The space $G/\mathbb{T}$ is the upper half-plane H, and the action of G preserves the Poincaré metric and area element on H. Corresponding to the 1-dimensional representations $z \mapsto z^p$ of $\mathbb{T}$ (with $p \in \mathbb{Z}$) we have the spaces $\Omega_{\text{hol}}^{p/2}$ of square-summable holomorphic $\frac{p}{2}$-forms $f(z)(dz)^{p/2}$ on H. The fractional powers $(dz)^{p/2}$ make sense, as under the Möbius transformation induced by

$$\begin{pmatrix} a & b \\ c & d \end{pmatrix}^{-1}$$

in $SL_2\mathbb{R}$ we have $dz \mapsto (cz + d)^{-2} dz$ and

$$(dz)^{p/2} \longmapsto (cz + d)^{-p} (dz)^{p/2}.$$

To analyse these representations further — and in particular to prove they are irreducible — we consider the action of the compact subgroup $\mathbb{T}$. It is best to replace $SL_2\mathbb{R}$ by $SU_{1,1}$, and hence the upper half-plane H by the unit disc D, on which $\mathbb{T}$ acts by rotation. Then an element of $\Omega_{\text{hol}}^{p/2}$ has a Taylor expansion

$$f(z)(dz)^{p/2} = \sum_{n \geq 0} a_n z^n (dz)^{p/2}. \tag{16.1}$$

Because $u \in \mathbb{T}$ acts on $z^n(dz)^{p/2}$ by multiplication by $u^{n+p/2}$, we can identify $(\Omega_{\text{hol}}^{p/2})^{\text{fin}}$, as a representation of $\mathbb{T}$, with the trigonometric polynomials

$$\sum_{m \geqslant p/2} a_m e^{im\theta}, \tag{16.2}$$

where m runs through the sequence $\frac{p}{2}, \frac{p}{2}+1, \frac{p}{2}+2, \ldots$. These form an irreducible representation of the Lie algebra, generated by the lowest-weight vector $e^{ip\theta/2}$.

The invariant norm of $f = f(z)(dz)^{p/2}$ is

$$\| f \|^2 = \int_D |f(z)|^2 (1-|z|^2)^{p-2} |dz d\bar{z}| \tag{16.3}$$

if $p > 1$, as the invariant Poincaré area element on D is

$$(1-|z|^2)^{-2} |dz d\bar{z}|.$$

In terms of the expansion (16.1) this means that

$$\| f \|^2 = \sum_{n \geqslant 0} K_n^{(p)} |a_n|^2,$$

where for each p, $\{K_n^{(p)}\}$ is a sequence of positive numbers which is $O(1/n^{p-1})$ as $n \to \infty$.

If $p \leqslant 1$ there are no non-zero square-summable holomorphic $(p/2)$-forms. But the case $p = 1$ is borderline: one can define a unitary representation $\Omega_{\text{hol}}^{\frac{1}{2}}$ by completing the holomorphic $\frac{1}{2}$-forms on the closed disc with respect to the invariant norm

$$\| f \|^2 = \frac{1}{2\pi} \int_0^{2\pi} |f(e^{i\theta})|^2 d\theta = \sum |a_n|^2,$$

which is a renormalization of the divergent expression (16.3).

The complex-conjugate representations $\bar{\Omega}_{\text{hol}}^{p/2}$ for $p \geqslant 1$ form another discrete series.

From (16.2) we see that the discrete series representations $\Omega_{\text{hol}}^{p/2}$ are roughly "half the size" of the principal series representations $E_{p,\epsilon}$. In fact if $\epsilon(p)$ is the parity of p then $\Omega_{\text{hol}}^{p/2}$ is a closed invariant subspace of the non-unitary representation $E_{p,\epsilon(p)}$, which is the space $\Omega^{p/2}$ of all $\frac{p}{2}$-forms on S^1. We have

$$\Omega_{\text{hol}}^{p/2} \oplus \bar{\Omega}_{\text{hol}}^{p/2} = \Omega_{(0)}^{p/2} \subset \Omega^{p/2},$$

and (16.2) shows that the quotient representation $V = \Omega^{p/2}/\Omega^{p/2}_{(0)}$ is $(p-1)$-dimensional, spanned by $e^{im\theta}$ with

$$-\frac{r}{2} < m < \frac{r}{2}.$$

But $\Omega^{p/2}$ does not decompose as a sum

$$\Omega^{p/2}_{\text{hol}} \oplus \bar{\Omega}^{p/2}_{\text{hol}} \oplus V.$$

The simplest case to understand is when $p = 2$. Then V is the trivial 1-dimensional representation of G, and $\Omega^1_{(0)}$ is the kernel of the equivariant map $\Omega^1 \to \mathbb{C}$ given by integration over S^1. But there is no G-invariant 1-form on S^1, and so the trivial representation does not occur in Ω^1.

To treat the general case we observe that G has a natural non-unitary irreducible representation on the $(q+1)$-dimensional space $P_q \cong S^q(\mathbb{C}^2)$ of homogeneous polynomials of degree q in two variables (u,v). The representation P_q is a subrepresentation of $\Omega^{-q/2} = E_{-q,\epsilon(q)}$ by the map

$$\vartheta(u,v) \longmapsto \vartheta(z,1)(dz)^{-q/2}.$$

We saw that if $q = p-2$ the representation $\Omega^{-q/2}$ is dual to $\Omega^{p/2}$, and the annihilator of P_{p-2} is easily seen to be $\Omega^{p/2}_{(0)}$. Thus

$$\Omega^{p/2}/\Omega^{p/2}_{(0)} \cong P^*_{p-2}.$$

Actually $P^*_{p-2} \cong P_{p-2}$, because the G-invariant skew form on $\mathbb{C}^2$ induces an invariant bilinear form on $S^q(\mathbb{C}^2)$ for each q. But there is no $(p-2)$-dimensional invariant subspace in $\Omega^{p/2}$.

The two kinds of irreducible representations of $SL_2\mathbb{R}$ we have described correspond, roughly speaking, to the two kinds of so-called *Cartan subgroups* (see page 12) of $SL_2\mathbb{R}$. For the present purpose a Cartan subgroup can be defined as a subgroup A whose complexification is an algebraic maximal torus $\mathbb{C}^\times$ (see page 179) of the complexified group $SL_2\mathbb{C}$. There are two kinds, for we have either $A \cong \mathbb{R}^\times$ or $A \cong \mathbb{T}$.

That concludes our account of the representations of $SL_2\mathbb{R}$. There is another series of irreducible unitary represenatations, called the *complementary series*, but it is of measure zero for the Plancherel measure.

A very careful and elementary account of all this material can be found in [Gelfand et al. Vol. 5].

17
The Heisenberg group, the metaplectic representation, and the spin representation

A basic problem in quantum mechanics is to find self-adjoint operators P and Q acting on a Hilbert space $\mathcal{H}$ such that

$$PQ - QP = -i \tag{17.1}$$

The problem, however, is not very well-posed, as P and Q are necessarily unbounded operators, and cannot be everywhere defined in $\mathcal{H}$. Weyl pointed out that it is better to think of the 1-parameter groups $\{e^{iaP}\}$ and $\{e^{ibQ}\}$ of unitary operators they generate. These operators are defined everywhere in $\mathcal{H}$ and satisfy

$$e^{iaP} e^{ibQ} = e^{-iab} e^{ibQ} e^{iaP}, \tag{17.2}$$

and so generate a 3-dimensional group H, which is called the *Heisenberg group*. Looking for unitary representations of H is a better way of formulating the basic problem (17.1).

We have met the group H in Chapter 1 as N/Z, and also as the group of unitary operators in the Hilbert space $\mathcal{H} = L^2(\mathbb{R})$ generated by translations T_a and multiplications M_b by the functions e^{ibx} on $\mathbb{R}$. These operators T_a and M_b satisfy the relations (17.2).

A more symmetrical description of H is as a central extension

$$\mathbb{T} \longrightarrow H \longrightarrow \mathbb{R}^2,$$

with the circle-group $\mathbb{T}$ as a normal subgroup, and $H/\mathbb{T} \cong \mathbb{R}^2$. We can write the elements of H as $u\exp(\xi)$ with $u \in \mathbb{T}$ and $\xi \in \mathbb{R}^2$. The group-law is

$$u\exp(\xi).v\exp(\eta) = uve^{iS(\xi,\eta)} \exp(\xi + \eta),$$

17 The Heisenberg group

where S is the skew bilinear form on $\mathbb{R}^2$ given by

$$S(\xi,\eta) = \xi_1\eta_2 - \xi_2\eta_1.$$

We are interested only in unitary representations of H in which the centre $\mathbb{T}$ acts by scalar multiplication. Stone and von Neuman proved that the only such representation is the one we already know on $\mathcal{H} = L^2(\mathbb{R})$. Nevertheless, there are other interesting ways of describing it. If we define

$$a = \frac{1}{\sqrt{2}}(P + iQ) \quad \text{and} \quad a^* = \frac{1}{\sqrt{2}}(P - iQ)$$

the relation (17.1) becomes

$$[a^*, a] = 1. \tag{17.3}$$

We observe that a^* acts on $\mathcal{H}$ as the operator

$$-\frac{i}{\sqrt{2}}(\frac{d}{dx} + x),$$

and annihilates the vector $\Omega = e^{-\frac{1}{2}x^2}$. In fact the elements $\{a^n\Omega\}_{n\geqslant 0}$ form an orthogonal basis for $\mathcal{H}$. Thus $\mathcal{H}$ contains a dense subspace $\mathcal{H}^{\text{fin}}$ which is isomorphic to the polynomial ring $\mathbb{C}[a]$, or, better, to the symmetric algebra $S(\mathbb{C})$. (Abstractly, what we have done in writing $a = \frac{1}{\sqrt{2}}(P+iQ)$ is choose a complex structure on the space $\mathbb{R}^2$, compatible with its skew form, and thereby we have identified $\mathbb{R}^2$ with $\mathbb{C}$.) When described as a completion of $\mathbb{C}[a]$ the Hilbert space $\mathcal{H}$ is known as the *oscillator representation*, for in quantum mechanics it describes the states of a simple harmonic oscillator. (The reason for the notation $\mathcal{H}^{\text{fin}}$ will appear.) Notice that a acts on $\mathbb{C}[a]$ by multiplication, and a^* acts as $\frac{d}{da}$.

The choice of Ω such that $a^*\Omega = 0$ is reminiscent of choosing a lowest-weight vector in a representation of $SL_2\mathbb{R}$, and this is no accident. For $SL_2\mathbb{R}$ acts on $\mathbb{R}^2$ preserving the skew form S which defines H, and so it acts on H as a group of automorphisms. As H has only one irreducible representation $\mathcal{H}$, Schur's lemma tells us we can find $T_g : \mathcal{H} \to \mathcal{H}$ for each $g \in SL_2\mathbb{R}$ such that

$$T_g \circ h = (gh) \circ T_g$$

for all $g \in SL_2\mathbb{R}$ and $h \in H$. The operators T_g are defined up to scalar multiplication, and define a projective representation of $SL_2\mathbb{R}$ on $\mathcal{H}$. We can find them without using the Stone-von Neumann theorem. The operators

$$\left\{\frac{i}{2}P^2,\ \frac{i}{2}(PQ+QP),\ \frac{i}{2}Q^2\right\} \tag{17.4}$$

are well-defined on $\mathcal{H}^{\text{fin}}$, and have the standard commutation relations of $sl_2\mathbb{R}$. They act on $L^2\mathbb{R}$ by

$$-\frac{i}{2}\left(\frac{d}{dx}\right)^2,\ x\frac{d}{dx}+\frac{1}{2},\ \frac{i}{2}x^2.$$

(These formulas should be compared with those of (10.7).) They generate an action of a double covering of $SL_2\mathbb{R}$ which is called the *metaplectic group* Mpl_2. Its action on $\mathcal{H}$ is the *metaplectic representation*.

We see that it is a double covering of $SL_2\mathbb{R}$ which acts on $\mathcal{H}$ as follows. The operator corresponding to the Lie algebra element ξ which generates the circle subgroup SO_2 in $SL_2\mathbb{R}$ is $\tilde{\xi} = iA$, where

$$A = \frac{1}{2}(P^2 + Q^2) = aa^* + \frac{1}{2},$$

which acts on $L^2(\mathbb{R})$ as

$$\frac{1}{2}\left\{-\left(\frac{d}{dx}\right)^2 + x^2\right\}$$

and on $\mathbb{C}[a]$ as $a\frac{d}{da} + \frac{1}{2}$. (In quantum mechanics this is the *Hamiltonian*, or energy operator, of the harmonic oscillator.) The eigenvectors of A are the basis vectors $\{a^n\Omega\}$, and the eigenvalues are $\{n+\frac{1}{2}\}$ for $n \geq 0$. So $\exp(2\pi\tilde{\xi}) = -1$, whereas $\exp(2\pi\xi) = 1$ in $SL_2\mathbb{R}$.

The group Mpl_2 is not a matrix group. For we know the finite dimensional representations of the Lie algebra $sl_2\mathbb{R}$ quite explicitly, (see page 30) and in any of them the operator $\tilde{\xi}$ representing ξ has eigenvalues of the form ni, where n is an integer. This means that $\exp(2\pi\tilde{\xi}) = 1$, and so a finite dimensional representation of Mpl_2 cannot be faithful.

The circle $\mathbb{T} = \{\exp\theta\xi\}_{0\leq\theta\leq 4\pi}$ in Mpl_2 which is the double covering of $SO_2 \subset SL_2\mathbb{R}$ is a maximal compact subgroup. Its action on $L^2\mathbb{R}$ by the operators $e^{i\theta A}$ is very beautiful. For most values of θ the operator is an integral operator with a kernel of the form

$$K_\theta e^{iQ_\theta(x,y)}$$

17 The Heisenberg group

where Q_θ is a quadratic form† on $\mathbb{R}^2$, and when $\theta = \pi/2$ we have

$$e^{\frac{1}{2}i\pi A} = e^{\frac{1}{4}i\pi}\mathscr{F},$$

where $\mathscr{F} : L^2(\mathbb{R}) \longrightarrow L^2(\mathbb{R})$ is the Fourier transform, which is of order four. But when θ is a multiple of π the kernel becomes a δ-function; thus

$$(e^{i\pi A}f)(x) = if(-x),$$

and

$$(e^{2\pi i A}f)(x) = -f(x).$$

Under the action of Mpl_2 the space $L^2(\mathbb{R})$ breaks up as $\mathscr{H}_+ \oplus \mathscr{H}_-$, where $\mathscr{H}_\pm$ are the even and odd functions on $\mathbb{R}$. These are irreducible: in the notation of Chapter 16 we have $\mathscr{H}_+ \cong \Omega^{\frac{1}{4}}_{\mathrm{hol}}$, the space of holomorphic $\frac{1}{4}$-forms $f(z)(dz)^{\frac{1}{4}}$ on the disc or upper half-plane. The map

$$\mathscr{H}_+ \longrightarrow \Omega^{\frac{1}{4}}_{\mathrm{hol}}(\text{upper half-plane})$$

takes $\psi \in L^2\mathbb{R}$ to $f(z)(dz)^{\frac{1}{4}}$, where

$$f(z) = \int \psi(x)e^{\frac{1}{2}izx^2}dx.$$

The metaplectic representation is of great importance not only in quantum mechanics but also in many other branches of mathematics, such as number theory (where the double covering is related to the law of quadratic reciprocity). It can be generalized in various ways. The most immediate is that a non-degenerate skew bilinear form S on $\mathbb{R}^{2n}$ defines a Heisenberg group H_{2n} which is an extension

$$\mathbb{T} \longrightarrow H_{2n} \longrightarrow \mathbb{R}^{2n}.$$

Studying the representations of H_{2n} is essentially the same as looking for operators $P_1, \ldots, P_n, Q_1, \ldots, Q_n$ such that

$$\left.\begin{array}{l}[P_k, P_l] = [Q_k, Q_l] = 0, \\ [P_k, Q_l] = -i\delta_{kl}.\end{array}\right\} \tag{17.5}$$

As when $n = 1$, the group H_{2n} has a unique faithful irreducible representation, on a Hilbert space $\mathscr{H}$ which is a completion of $\mathscr{H}^{\mathrm{fin}} = S(\mathbb{C}^n)$. The symplectic group $Sp_{2n}\mathbb{R}$ acts on H_{2n} as a group of automorphisms, and a double covering Mpl_{2n} of $Sp_{2n}\mathbb{R}$ acts on $\mathscr{H}$.

† In fact Q_θ is the generating function of the "contact transformation" $\exp(\theta\xi)$ of $\mathbb{R}^2$, in the sense of classical mechanics.

The spin representation

The metaplectic representation is an exact analogue of the *spin representation* of a double covering of SO_{2n}. If we replace the skew form S on $\mathbb{R}^{2n}$ by a symmetric one we can pose the problem, analogous to (17.1) and (17.5), of finding self-adjoint operators $\{P_i\}_{1 \leq i \leq 2n}$ on a Hilbert space $\mathcal{H}$ such that

$$P_i P_j + P_j P_i = \delta_{ij}.$$

In this case the operators P_i are automatically bounded (because $P_i^2 = \frac{1}{2}$), and they generate a finite dimensional algebra C_{2n}, which is called the *Clifford algebra*. It has dimension 2^{2n}, and turns out to be isomorphic to the algebra of all $2^n \times 2^n$ matrices. Looking for representations of C_{2n} corresponds to looking for representations of the Heisenberg group H_{2n}. There is a unique irreducible representation, and it can be realized on $\bigwedge(\mathbb{C}^n)$.

The elements $\frac{1}{2}(P_i P_j - P_j P_i)$ of C_{2n}, analogous to the elements (17.4), have the commutation relations of the Lie algebra SO_{2n}. They generate a representation of a double covering of SO_{2n} on $\bigwedge(\mathbb{C}^n)$. This is the spin representation.

Linear Algebraic Groups

I. G. Macdonald

Contents
Linear Algebraic Groups

Preface		135
Introduction		137
1	Affine algebraic varieties	139
2	Definition and elementary properties	146
Interlude		154
3	Projective algebraic varieties	157
4	Tangent spaces. Separability	162
5	Lie algebra of a linear algebraic group	166
6	Homogeneous spaces and quotients	172
7	Borel subgroups and maximal tori	177
8	Root structure	182

Preface

These notes follow the course of the lectures I gave at Lancaster, but contain rather more detail than it was possible to include in seven hours' lecturing time. Even so, since any adequate account of the theory of linear algebraic groups requires a book of two to three hundred pages, it is obvious that I have had to leave a lot out. I have attempted only to convey the flavour of the subject, though I am painfully aware that the experts (who ought not to be reading this anyway) may well disagree with my choice of what to put in and what to leave out, what to prove and what not to prove.

In preparing these notes I have relied heavily on the books of Borel [B], Humphreys [H] and Springer [S], to which these notes may perhaps serve as an introduction. At the end, under **Notes and References**, I give references to these books for the proofs of theorems not proved in the text.

[B] A. Borel, *Linear Algebraic Groups* (Math. Lecture Note Series, W.A. Benjamin, Inc. New York, 1969).
[H] James E. Humphreys, *Linear Algebraic Groups* (Graduate Texts in Mathematics, Springer-Verlag, New York, 1975).
[S] T. A. Springer, *Linear Algebraic Groups* (Birkhaüser, Boston, 1981)

Introduction

Let K be a field (in fact K will always be algebraically closed, for example the field of complex numbers, or an algebraic closure of a finite field), and let $M_n(K)$ denote the space of all $n \times n$ matrices $x = (x_{ij})$ with entries x_{ij} in K. The determinant of $x \in M_n(K)$ is denoted by $\det x$, and the transpose of x by x^t.

Let
$$GL_n(K) = \{x \in M_n(K) : \det x \neq 0\},$$

the *general linear group*;

$$SL_n(K) = \{x \in M_n(K) : \det x = 1\},$$

the *special linear group*;

$$O_n(K) = \{x \in M_n(K) : x^t x = 1_n\},$$

the *orthogonal group*;

$$SO_n(K) = O_n(K) \cap SL_n(K),$$

the *special orthogonal group*;

$$Sp_{2n}(K) = \{x \in M_{2n}(K) : x^t j x = j\}$$

(where $j = \begin{pmatrix} 0 & 1_n \\ -1_n & 0 \end{pmatrix}$), the *symplectic group*.

All of these are examples of *linear algebraic groups*. Each of them is (a) a group, the group operation in each case being matrix multiplication; and (b) an algebraic variety, that is to say defined by polynomial equations in the matrix coordinates x_{ij}. For example, $O_n(K)$ is defined by the equations

$$\sum_{j=1}^{n} x_{ij} x_{kj} = \delta_{ik} \qquad (1 \leq i, k \leq n)$$

where $\delta_{ik} = 1$ or 0 according as $i = k$ or $i \neq k$. (The general linear group $GL_n(K)$ appears to be an exception: however, if we introduce another coordinate z, we may regard $GL_n(K)$ as the set of points $(x_{ij}, z) \in M_n(K) \times K = K^{n^2+1}$ that satisfy $z \det(x_{ij}) = 1$.)

Moreover, if x and y are elements of any one of these groups, the coordinates of xy (resp. of x^{-1}) are polynomial functions of the coordinates of x and y (resp. of x).

Some knowledge of algebraic geometry is an essential prerequisite to any study of linear algebraic groups, and in §1 we shall present the reader with just enough to get the subject started. Later sections will provide additional doses as and when required.

1
Affine algebraic varieties

Let K be an algebraically closed field, of any characteristic (for example, the field $\mathbb{C}$ of complex numbers, or an algebraic closure of a finite field). Let

$$A = A_n = K[t_1, \ldots, t_n]$$

where $t_1, \ldots, t_n$ are independent indeterminates over K. The elements of A are polynomials in the t_i with coefficients in K, and we may regard them as K-valued *functions* on the affine space $K^n = K \times \ldots \times K$ (n factors): if $f \in A$ and $x = (x_1, \ldots, x_n) \in K^n$, then $f(x) = f(x_1, \ldots, x_n) \in K$ is the result of substituting x_i for t_i ($1 \leq i \leq n$) in f. In particular, $t_i(x) = x_i$, so that t_i is the ith *coordinate function* on K^n.

An *algebraic set* in K^n is traditionally defined by a finite set of polynomial equations, say

$$(*) \qquad f_i(x) = 0 \qquad (1 \leq i \leq r)$$

where the f_i are in A. More precisely, it is the set (perhaps empty) of all $x \in K^n$ satisfying the equations $(*)$. However, in making the definitions there is no need to restrict the number of equations to be finite; so we start with *any* subset $S \subset A$ and define

$$V(S) = \{x \in K^n : f(x) = 0 \text{ for all } f \in S\}$$

so that $V(S)$ is the set of all points in K^n at which all the polynomials in S vanish.

This operation V has the following properties, in which S, S_1, S_2, S_i are arbitrary subsets of A. The proofs are straightforward.

(1.1) (i) $V(S_1) \cup V(S_2) = V(S_1 S_2)$, where $S_1 S_2$ is the set of products $f_1 f_2$ with $f_1 \in S_1$ and $f_2 \in S_2$.
(ii) $\bigcap_{i \in J} V(S_i) = V(\bigcup_{i \in J} S_i)$ for any index set J (finite or infinite).
(iii) $V(A) = \emptyset$, $V(\emptyset) = K^n$.
(iv) $S_1 \subset S_2 \Rightarrow V(S_1) \supset V(S_2)$.
(v) Let $\mathfrak{a}$ be the ideal in A generated by S; then $V(S) = V(\mathfrak{a})$.

(*Remark.* The ring A is Noetherian (Hilbert's basis theorem), hence the ideal $\mathfrak{a}$ in (v) above has a finite basis, say S_0, and $V(S) = V(\mathfrak{a}) = V(S_0)$. Thus there would be no loss of generality in assuming that the set of equations defining V is finite.)

Consider in particular the statements (i)—(iii) above. They show that the sets $V(S)$, $S \subset A$, satisfy the axioms for *closed* sets in a topological space. The resulting topology on K^n is called the *Zariski topology*, and the induced topology on an algebraic set $X \subset K^n$ is the *Zariski topology of X*. Thus the open sets in K^n are the complements of the algebraic sets $V(S)$. Intuitively, the non-empty open sets are very large: for example in the affine line ($n = 1$) they are the complements of finite subsets of K, and so in particular any two non-empty open sets in the affine line always intersect.

By (1.1)(v) we may restrict our attention to subsets of A that are ideals, and the operation V takes ideals in A to (certain) subsets of K^n. In the opposite direction, let E be any subset of K^n and define

$$I(E) = \{f \in A : f(x) = 0, \text{ all } x \in E\}.$$

Clearly $I(E)$ is an ideal in A, and so the operation I takes subsets of K^n to ideals in A. It is not difficult to check that if $E \subset K^n$ then

$$V(I(E)) = \overline{E},$$

the closure of E in the Zariski topology.

Next, if $\mathfrak{a}$ is an ideal of A we have

$$I(V(\mathfrak{a})) = r(\mathfrak{a})$$

where $r(\mathfrak{a})$ is the *radical* of $\mathfrak{a}$, namely the set of all $f \in A$ some power of which lies in $\mathfrak{a}$. This is not an obvious result: it is a famous theorem of Hilbert (the Nullstellensatz).

Now $r(\mathfrak{a})$ is a radical ideal (i.e., equal to its radical). It follows from what we have said that the operations V and I are mutually inverse

1 Affine algebraic varieties

order-reversing bijections between the set of algebraic subsets of K^n and the set of radical ideals in A_n:

(1.2) $\qquad$ (alg. subsets $X \subset K^n$) $\underset{V}{\overset{I}{\rightleftarrows}}$ (radical ideals $\mathfrak{a} \subset A_n$).

As remarked earlier, the ring $A_n = K[t_1,\ldots,t_n]$ is Noetherian, so that the ideals in A_n satisfy the ascending chain condition. From (1.2) it follows that the closed sets in K^n (i.e., the algebraic subsets) satisfy the descending chain condition.

In general, a topological space X is said to be *Noetherian* if it satisfies the descending chain condition on closed subsets, that is to say if every strictly decreasing sequence of closed sets in X:

$$X_1 \supset X_2 \supset \ldots \supset X_r \supset \ldots$$

(with strict inclusions at each stage) is finite.

(1.3) *Let X be a Noetherian topological space. Then*
(i) *X is quasi-compact (i.e., every open covering of X has a finite subcovering).*
(ii) *Every closed subset of X (with the induced topology) is Noetherian.*

Next, a topological space X is *irreducible* if it is *not* the union of two proper closed subsets, or equivalently if any two non-empty open subsets of X intersect. Irreducible implies connected, but not conversely (the subvariety $x_1 x_2 = 0$ of K^2 is connected but not irreducible).

(1.4) *Let E be a subset of a topological space X.*
(i) *E (with the induced topology) is irreducible if and only if $\overline{E}$ is irreducible.*
(ii) *If $f : X \to Y$ is a continuous mapping and E is irreducible then $f(E)$ is irreducible.*

(1.5) *Let X be a Noetherian topological space. Then X is the union of finitely many irreducible closed subsets, say*

$$X = X_1 \cup \ldots \cup X_r.$$

If $X_i \not\subset X_j$ for all pairs $i \neq j$, then this decomposition of X is unique, and the X_i are the maximal irreducible subsets of X (relative to inclusion).

The closed subsets X_i in (1.5) are the *irreducible components* of X. In particular, if $X \subset K^n$ is an algebraic set, then X (with the induced topology) is Noetherian by (1.3)(ii) (since K^n is Noetherian), and (1.5) applies to X.

Instead of concentrating attention on the ideal $\mathfrak{a} = I(X)$, where $X \subset K^n$ is an algebraic set, it is better to factor it out and consider the K-algebra $A_n/\mathfrak{a}$. This is called the *affine algebra* (or *coordinate ring*) of X, and it may be described as follows. A function φ on X with values in K is said to be *regular* if it is the restriction to X of some $f \in A_n$. The regular functions on X form a K-algebra $K[X]$ under pointwise addition and multiplication, and restriction to X defines a surjective homomorphism of A_n onto $K[X]$ whose kernel is precisely $I(X) = \mathfrak{a}$, so that $K[X] \cong A_n/\mathfrak{a}$. As a K-algebra, $K[X]$ is generated by the restrictions to X of the coordinate functions t_i ($1 \leqslant i \leqslant n$).

Each point $x \in X$ determines a K-algebra homomorphism

$$\epsilon_x : K[X] \to K,$$

namely $f \mapsto f(x)$ (evaluation at x). Thus we have a mapping

$$X \to \mathrm{Hom}_{K-\mathrm{alg.}}(K[X], K)$$

which in fact is a *bijection*. Thus X can be reconstructed from its affine algebra $K[X]$, and the closed subsets of X are the sets

$$V_X(S) = \{x \in X : f(x) = 0 \text{ for all } f \in S\}$$

for all subsets S of $K[X]$. Abstractly, then, X is a topological space carrying a ring $K[X]$ of K-valued functions on X. The pair $(X, K[X])$ is an *affine algebraic variety*, and every finitely generated commutative K-algebra with no nilpotent elements $\neq 0$ occurs as $K[X]$. In practice we shall habitually drop $K[X]$ from the notation, and speak of X as an affine algebraic variety.

(1.6) *Let X be an affine algebraic variety. Then X is irreducible if and only if $K[X]$ is an integral domain.*

In particular, affine space K^n is irreducible.

If X is irreducible, the integral domain $K[X]$ has a field of fractions, denoted by $K(X)$ and called the field of rational functions or *function field* of X. It is a finitely generated field extension of K.

Morphisms

First of all, a mapping $f : K^n \to K^m$ is a *morphism* (or regular mapping) if there exist polynomials $f_1, \ldots, f_m$ in $A_n = K[t_1, \ldots, t_n]$ such that $f(x) = (f_1(x), \ldots, f_m(x))$ for all $x \in K^n$.

More generally, if $X \subset K^n$ and $Y \subset K^m$ are affine algebraic varieties, a mapping $\varphi : X \to Y$ is a *morphism* if $\varphi = f|X$ for some morphism $f : K^n \to K^m$ as above. Equivalently, φ is a morphism if and only if $g \circ \varphi$ is a regular function on X whenever g is a regular function on Y. It follows that φ determines a K-algebra homomorphism

$$\varphi^* : K[Y] \to K[X],$$

namely $\varphi^*(g) = g \circ \varphi$. It is easily verified that if S is any subset of $K[Y]$, then $\varphi^{-1}(V_Y(S)) = V_X(\varphi^*S)$, which shows that φ is a continuous mapping.

Conversely, every K-algebra homomorphism of $K[Y]$ into $K[X]$ is equal to φ^* for some morphism $\varphi : X \to Y$.

If $\varphi : X \to Y$ and $\psi : Y \to Z$ are morphisms, then $\psi \circ \varphi : X \to Z$ is a morphism, and $(\psi \circ \varphi)^* = \varphi^* \circ \psi^*$.

(1.7) Let $\varphi : X \to Y$ be a morphism of affine algebraic varieties.
(i) φ^* *is injective if and only if* $\overline{\varphi(X)} = Y$ *(one says that φ is dominant)*.
(ii) φ^* *is surjective if and only if φ is an isomorphism of X onto a closed subvariety of Y*.

An *isomorphism* of affine algebraic varieties is a bijective morphism $\varphi : X \to Y$ such that $\varphi^{-1} : Y \to X$ is also a morphism: equivalently, φ is an isomorphism if and only if $\varphi^* : K[Y] \to K[X]$ is an isomorphism of K-algebras. It should be remarked that it can happen that $\varphi : X \to Y$ is bijective, and indeed a homeomorphism, but *not* an isomorphism.

(1.8) *Example.* Let $X = Y = K$ (the affine line), and $\varphi(x) = x^p$, where $p > 0$ is the characteristic of K. Then φ is a homeomorphism, but $\varphi^* : K[t] \to K[t]$ is the homomorphism defined by $t \mapsto t^p$, and therefore φ is not an isomorphism.

Products

Let $X \subset K^n$, $Y \subset K^m$ be affine algebraic varieties. Then $X \times Y \subset K^n \times K^m = K^{n+m}$ is a closed subset (i.e., subvariety) of K^{n+m}. Let $f \in K[X]$, $g \in K[Y]$, and define a function $f \cdot g$ on $X \times Y$ by $(f \cdot g)(x, y) = f(x)g(y)$. Then $f \cdot g$ is a regular function on $X \times Y$, and the mapping $(f, g) \mapsto f \cdot g$

from $K[X] \times K[Y]$ to $K[X \times Y]$ is K-bilinear. Hence by the universal property of tensor products it gives rise to a mapping

$$\alpha : K[X] \otimes_K K[Y] \to K[X \times Y]$$

such that $\alpha(f \otimes g) = f \cdot g$. This mapping α is in fact an isomorphism of K-algebras, so that

(1.9) $K[X \times Y] \cong K[X] \otimes_K K[Y]$.

Next we have

(1.10) *If X, Y are irreducible affine varieties, then $X \times Y$ is irreducible.*

It should be remarked that if X, Y are affine varieties, the Zariski topology on $X \times Y$ is in general *finer* (i.e., has more open sets) than the product of the Zariski topologies on X and Y. (Consider the case $X = Y = K^1$.)

(1.11) *Let X be an affine algebraic variety. Then the diagonal*

$$\Delta_X = \{(x,x) : x \in X\}$$

is a closed subset of $X \times X$.

This is a sort of substitute for the Hausdorff axiom: if X is a topological space and $X \times X$ is given the product topology, then X is Hausdorff if and only if Δ_X is closed in $X \times X$.

Finally, if $\varphi : X \to Y$ and $\varphi' : X' \to Y'$ are morphisms, then $\varphi \times \varphi' : X \times X' \to Y \times Y'$ is a morphism, and $(\varphi \times \varphi')^* = \varphi^* \otimes \varphi'^*$ (when $K[X \times X']$ and $K[Y \times Y']$ are identified with $K[X] \otimes K[X']$ and $K[Y] \otimes K[Y']$ respectively, via the isomorphism (1.9)).

The image of a morphism

If $\varphi : X \to Y$ is a morphism of affine algebraic varieties, the image $\varphi(X)$ of φ need not be either open or closed in Y. For example, if $X = Y = K^2$ and $\varphi(x_1, x_2) = (x_1 x_2, x_2)$, then the image of φ consists of the complement of the line $x_2 = 0$, together with the point $(0,0)$, hence is the union of an open set and a closed set.

A subset E of a topological space X is *locally closed* if each $x \in E$ has an open neighbourhood U_x in X such that $E \cap U_x$ is closed in U_x. Equivalently, E is locally closed if and only if E is the intersection of an

1 Affine algebraic varieties

open set and a closed set, or again if and only if E is open in its closure $\overline{E}$.

Next, $E \subset X$ is *constructible* if it is a finite union of locally closed subsets. If E is constructible and not empty, then E contains a non-empty open subset of $\overline{E}$.

If E and F are constructible, so also are $E \cup F$, $E \cap F$ and the complement $X - E$. The constructible subsets of X are precisely the elements of the Boolean algebra generated by the open subsets of X.

(1.12) *Let $\varphi : X \to Y$ be a morphism of affine algebraic varieties, E a constructible subset of X. Then $\varphi(E)$ is a constructible subset of Y.*

The proof of (1.12) can be reduced to showing that

(1.12′) *If X is an irreducible affine variety and $\varphi : X \to Y$ a dominant morphism (1.7), then $\varphi(X)$ contains a non-empty open subset of Y.*

This in turn rests on the following proposition from commutative algebra:

Let $A \subset B$ be integral domains such that B is finitely generated as an A-algebra, and let K be an algebraically closed field. Let $b \in B$, $b \neq 0$. Then there exists $a \neq 0$ in A with the following property: every homomorphism $\epsilon : A \to K$ such that $\epsilon(a) \neq 0$ extends to a homomorphism $\epsilon' : B \to K$ such that $\epsilon'(b) \neq 0$.

Dimension

Let X be an irreducible affine algebraic variety, $K(X)$ its function field. The *dimension* $\dim X$ of X is defined to be the transcendence degree of $K(X)$ over K, that is to say the maximum number of elements of $K(X)$ that are algebraically independent over K. For example, when $X = K^n$ we have $K(X) = K(t_1, \ldots, t_n)$ and hence $\dim K^n = n$.

If now X is reducible, with irreducible components $X_1, \ldots, X_r$ (1.5), we define $\dim X$ to be the maximum of the dimensions of the components X_i. In particular, $\dim X = 0$ if and only if X is a finite set.

(1.13) *Let X be an irreducible affine algebraic variety, Y an irreducible closed subvariety of X. If $Y \neq X$ then $\dim Y < \dim X$.*

(1.14) *Let X, Y be irreducible affine varieties. Then $\dim(X \times Y) = \dim X + \dim Y$.*

2
Linear algebraic groups: definition and elementary properties

A *linear algebraic group* G is (1) an affine algebraic variety and (2) a group, such that multiplication

$$\mu : G \times G \to G \qquad (\mu(x,y) = xy)$$

and inversion

$$\iota : G \to G \qquad (\iota(x) = x^{-1})$$

are morphisms of affine algebraic varieties. (The reason for the adjective "linear" (rather than "affine") will become plain later, in (2.5).)

If G, H are linear algebraic groups, so is their direct product $G \times H$.

A mapping $\varphi : G \to H$ is a *homomorphism of algebraic groups* if φ is (1) a morphism of affine varieties and (2) a homomorphism of groups.

Let G be a linear algebraic group, $A = K[G]$ its affine algebra. The K-algebra A carries a "comultiplication"

$$\mu^* : A \to A \otimes_K A$$

which makes it into a (coassociative) Hopf algebra with antipode $\iota^* : A \to A$. From this point of view, the study of linear algebraic groups is equivalent to the study of a certain class of Hopf algebras; but we shall not pursue this line of thought.

Examples

1. $G = K$ with addition as group operation: $\mu(x,y) = x+y$, $\iota(x) = -x$. We have $K[G] = K[t]$ with

$$\mu^* : K[t] \to K[t] \otimes K[t] = K[t_1, t_2]$$

given by $\mu^*(t) = t_1 + t_2$, and

$$\iota^* : K[t] \to K[t]$$

2 Linear algebraic groups: definition and elementary properties

given by $\iota^*(t) = -t$. G is the *additive group*, usually denoted by $\mathbb{G}_a$. Clearly $\dim \mathbb{G}_a = 1$.

2. $G = K^* = K - \{0\}$ with multiplication as group operation: $\mu(x,y) = xy$, $\iota(x) = x^{-1}$. We may identify G with the closed subvariety of K^2 given by the equation $x_1 x_2 = 1$, so that $K[G] = K[t, t^{-1}]$ with

$$\mu^* : K[t, t^{-1}] \to K[t_1, t_2, t_1^{-1}, t_2^{-1}]$$

given by $\mu^*(t) = t_1 t_2$, and $\iota^*(t) = t^{-1}$. G is the *multiplicative group*, usually denoted by $\mathbb{G}_m$ (or GL_1).

3. $G = GL_n(K)$ is the group of nonsingular $n \times n$ matrices over K (i.e., matrices x such that $\det x \neq 0$). We identify G with the closed subset of points $(x, y) \in K^{n^2} \times K = K^{n^2+1}$ such that $y \det(x) = 1$. Thus

$$K[G] = K[t_{ij} (1 \leq i, j \leq n), d^{-1}]$$

where $d = \det(t_{ij})$, and

$$\mu^*(t_{ij}) = \sum_{k=1}^{n} t_{ik} \otimes t_{kj}.$$

$GL_n(K)$ is irreducible, of dimension n^2.

4. Any Zariski-closed subgroup of $GL_n(K)$ is a linear algebraic group. Thus, apart from SL_n, O_n, SO_n, Sp_{2n} mentioned in the introduction, the following are linear algebraic groups:

(a) any finite subgroup of $GL_n(K)$;

(b) D_n, the group of nonsingular diagonal matrices

$$\begin{pmatrix} x_1 & & & \\ & x_2 & & \\ & & \ddots & \\ & & & x_n \end{pmatrix},$$

isomorphic to $\mathbb{G}_m^n$;

(c) B_n, the group of upper triangular matrices $x = (x_{ij}) \in GL_n(K)$ such that $x_{ij} = 0$ if $i > j$;

(d) U_n, the group of upper unipotent matrices ($x_{ij} = 0$ if $i > j$; $x_{ii} = 1$ ($1 \leq i \leq n$)).

(2.1) *Let G be a linear algebraic group. Then G has a unique irreducible component G_0 containing the identity element e, and G_0 is a closed normal subgroup of finite index in G. The irreducible components of G are also the connected components of G, and are the cosets of G_0 in G.*

As a general remark before coming to the proof of (2.1), if $x \in G$ the mappings
$$\lambda_x : g \mapsto xg, \quad \rho_x : g \mapsto gx$$
are *automorphisms* of the algebraic variety G, and in particular are homeomorphisms of the underlying topological space. For example, λ_x is the composition $g \mapsto (x,g) \mapsto \mu(x,g) = xg$, hence is a morphism of affine varieties, with inverse $\lambda_{x^{-1}}$.

Proof of (2.1). Let X, Y be irreducible components of G containing e. Then $XY = \mu(X \times Y)$ is irreducible, by (1.4) and (1.9). But XY contains X and Y, hence (as X, Y are maximal irreducible subsets of G) $X = XY = Y$.

It follows that $XX = X$, whence X is closed under multiplication; also $X^{-1} = \iota(X)$ is irreducible and contains e, so that $X^{-1} \subset X$. Hence $X = G_0$ is a subgroup of G, and is closed (because irreducible components are closed). Again, if $x \in G$ then $xG_0x^{-1} = \lambda_x \rho_{x^{-1}}(G_0)$ is an irreducible component of G containing e, hence is equal to G_0. So G_0 is a closed normal subgroup of G.

By translation, the unique irreducible component of G containing a given $x \in G$ is $\lambda_x G_0 = xG_0$. It follows that the irreducible components of G are the cosets of G_0 in G, and so by (1.5) G_0 has finite index in G. Each coset xG_0 is closed and therefore G_0, being the complement of the union of the cosets $xG_0 \neq G_0$, is *open* in G.

Finally, G_0 is connected (because irreducible), and since it is both open and closed in G it is the connected component of e in G.

From (2.1) it follows that for a linear algebraic group, irreducibility is equivalent to connectedness. It is customary to speak of a *connected* (rather than irreducible) algebraic group. The groups $\mathbb{G}_a$, $\mathbb{G}_m$, GL_n, SL_n, D_n, B_n, U_n are all connected; the group O_n is not (if char. $K \neq 2$). The groups SO_n and Sp_{2n} are in fact connected (but this is not obvious at this stage).

(2.2) Let G be a linear algebraic group, H a subgroup of G.
(i) $\overline{H}$ is a subgroup of G.
(ii) If H is constructible then it is closed in G.

Proof. (i) Let $x \in H$, so that $H = xH = \lambda_x H$. Take closures: $\overline{H} = \overline{\lambda_x H} = \lambda_x \overline{H} = x\overline{H}$, whence $H\overline{H} \subset \overline{H}$ and therefore $Hy \subset \overline{H}$ for all $y \in \overline{H}$. Take closures again: $\overline{H}y = \overline{Hy} \subset \overline{H}$, whence $\overline{H}.\overline{H} \subset \overline{H}$. Also $\overline{H}^{-1} = \overline{H^{-1}} = \overline{H}$, and hence $\overline{H}$ is a group.

2 Linear algebraic groups: definition and elementary properties

(ii) Since H is constructible it contains a non-empty open subset U of $\bar{H}$. But then H is a union of translates of U, hence is open in $\bar{H}$. Hence the cosets of H in $\bar{H}$ are open in $\bar{H}$, and therefore H, being the complement in $\bar{H}$ of the union of the cosets $xH \neq H$, is closed in $\bar{H}$; so $H = \bar{H}$ is closed in G.

(2.3) *Let $\varphi : G \to H$ be a homomorphism of linear algebraic groups. Then*
(i) *$\operatorname{Ker} \varphi$ is a closed subgroup of G.*
(ii) *$\operatorname{Im} \varphi = \varphi(G)$ is a closed subgroup of H.*
(iii) *$\varphi(G)_0 = \varphi(G_0)$.*

Proof. (i) φ is continuous, hence $\operatorname{Ker} \varphi = \varphi^{-1}(\{e\})$ is closed in G (and is of course a subgroup of G).
(ii) $\varphi(G)$ is constructible (1.12), and is a subgroup of H, hence is closed in H by (2.2).
(iii) $\varphi(G_0)$ is irreducible by (1.4)(ii) and is a closed subgroup of finite index in $\varphi(G)$, hence is open in $\varphi(G)$ and therefore is the identity component $\varphi(G)_0$ of $\varphi(G)$.

The next fact to be established is that every linear algebraic group G is isomorphic (as an algebraic group) to a closed subgroup of some $GL_n(K)$. (Thus the general linear groups $GL_n(K)$ play the same role in the theory of linear algebraic groups as the symmetric groups do in the theory of finite groups.)

Let G be a linear algebraic group, $A = K[G]$ its affine algebra. G acts on A by right translations:

$$(\rho(x)f)(y) = f(yx),$$

and by left translations:

$$(\lambda(x)f)(y) = f(x^{-1}y),$$

for $f \in A$ and $x, y \in G$. In what follows we shall use right translations rather than left translations. This is purely a matter of choice. In (2.4) below, therefore, "G-stable" means "stable under $\rho(x)$ for all $x \in G$".

(2.4) *Let V be a finite-dimensional K-vector subspace of $A = K[G]$. Then*
(i) *V is contained in a finite-dimensional G-stable vector subspace of A.*
(ii) *V is G-stable if and only if $\mu^*(V) \subset V \otimes_K A$.*

Proof. (i) It is enough to consider the case where $\dim V = 1$, say $V = Kf$, $f \in A$. Let

$$\mu^*(f) = \sum_{i=1}^{n} f_i \otimes g_i$$

say, with $f_i, g_i \in A$. Then for $x, y \in G$ we have

$$\begin{aligned}(\rho(x)f)(y) &= f(yx) = f(\mu(y,x)) \\ &= (\mu^*f)(y,x) \\ &= \sum_i f_i(y)g_i(x)\end{aligned}$$

and therefore

$$\rho(x)f = \sum_i g_i(x)f_i \in \sum_1^n Kf_i.$$

Thus the G-orbit of f is contained in the subspace of A spanned by $f_1, \ldots, f_n$, and therefore spans a finite-dimensional K-vector subspace of A.

(ii) Let $v_1, \ldots, v_n$ be a K-basis of V, and adjoin (infinitely many) elements $w_\alpha \in A$ to obtain a K-basis of A. Let $f \in V$, then μ^*f can be written in the form

$$\mu^*f = \sum_{i=1}^{n} v_i \otimes v'_i + \sum_\alpha w_\alpha \otimes w'_\alpha$$

for suitable $v'_i, w'_\alpha \in A$ (and almost all $w'_\alpha = 0$). If now $x \in G$ it follows as above that

$$\rho(x)f = \sum_i v'_i(x)v_i + \sum_\alpha w'_\alpha(x)w_\alpha$$

so that $\rho(x)f \in V$ if and only if $w'_\alpha(x) = 0$ for all α. Hence

$$\begin{aligned}\rho(x)f \in V \text{ for all } x \in G &\Leftrightarrow w'_\alpha = 0 \text{ for all } \alpha \\ &\Leftrightarrow \mu^*f = \sum_i v_i \otimes v'_i \\ &\Leftrightarrow \mu^*f \in V \otimes_K A.\end{aligned}$$

(2.5) *Let G be a linear algebraic group. Then G is isomorphic (as an algebraic group) to a closed subgroup of $GL_n(K)$ for some $n \geq 1$.*

Proof. The affine algebra $A = K[G]$ is a finitely-generated K-algebra, say $A = K[v_1, \ldots, v_n]$. By (2.4)(i) we may assume that the subspace $V = \sum Kv_i$ of A is G-stable (G acting by right translations), and that the

2 Linear algebraic groups: definition and elementary properties 151

v_i are linearly independent over K. By (2.4)(ii) we have $\mu^*(V) \subset V \otimes_K A$, and hence equations

(1) $$\mu^*(v_j) = \sum_{i=1}^{n} v_i \otimes \varphi_{ij}$$

for suitable $\varphi_{ij} \in A$. From (1) it follows that

(2) $$\rho(x)v_j = \sum_{i=1}^{n} \varphi_{ij}(x)v_i$$

for all $x \in G$, and hence (since $\rho(xy) = \rho(x)\rho(y)$) that $\varphi : x \mapsto (\varphi_{ij}(x))$ is a homomorphism of algebraic groups mapping G into $GL_n(K)$. To show that φ is an isomorphism of G onto a closed subgroup of $GL_n(K)$, it is enough by (1.7) to check that φ^* is surjective. We have $K[GL_n] = K[t_{ij}(1 \leq i, j \leq n), d^{-1}]$, where t_{ij} are the coordinate functions on $GL_n(K)$, and $d = \det(t_{ij})$; and $(\varphi^* t_{ij})(x) = t_{ij}(\varphi(x)) = \varphi_{ij}(x)$, so that $\varphi^*(t_{ij}) = \varphi_{ij}$. From (2) above we have

$$v_j(x) = (\rho(x)v_j)(e) = \sum_{i=1}^{n} \varphi_{ij}(x)v_i(e),$$

so that

$$v_j = \sum_{i} v_i(e)\varphi_{ij} = \varphi^*\left(\sum_{i} v_i(e)t_{ij}\right).$$

Hence each generator v_j of A lies in the image of φ^*, and so φ^* is surjective, as required.

Jordan decomposition

A matrix $x \in M_n(K)$ is said to be

semisimple if it is diagonalisable, i.e. if there exists $g \in GL_n(K)$ such that gxg^{-1} is a diagonal matrix;

nilpotent if $x^m = 0$ for some positive integer m, i.e. if the only eigenvalue of x is 0;

unipotent if $x - 1_n$ is nilpotent, i.e. if the only eigenvalue of x is 1.

(2.6) *Let $x, y \in M_n(K)$ commute ($xy = yx$).*
(i) *If x, y are semisimple then $x + y$ and xy are semisimple.*
(ii) *If x, y are nilpotent then $x + y$ and xy are nilpotent.*
(iii) *If x, y are unipotent then xy is unipotent.*

Proof. (i) follows from the fact that commuting semisimple matrices can be simultaneously diagonalized.

(ii) is true in any ring.

(iii) We have $x = 1 + a$ and $y = 1 + b$ where a, b are nilpotent and commute. Hence $xy = 1 + c$ where $c = a + b + ab$ is nilpotent by (ii).

(2.7) *Let $x \in GL_n(K)$. Then there exist x_s, $x_u \in GL_n(K)$ such that x_s is semisimple, x_u unipotent, and $x = x_s x_u = x_u x_s$. Moreover x_s and x_u are uniquely determined by these conditions.*

It follows from the definitions that a matrix $x \in GL_n(K)$ is semisimple or unipotent if and only if gxg^{-1} has the same property, for any $g \in G$. Hence to define x_s and x_u we may replace x by any conjugate of x in $GL_n(K)$. Thus we may replace x by its Jordan canonical form: there exists $g \in GL_n(K)$ such that the matrix gxg^{-1} is a diagonal sum of Jordan blocks

$$J_r(\lambda) = \begin{pmatrix} \lambda & 1 & & \\ & \lambda & 1 & \\ & & \ddots & 1 \\ & & & \lambda \end{pmatrix}$$

(r being the size of the block, and $\lambda \in K$ an eigenvalue of x). Thus it is enough to define x_s and x_u when x is a Jordan block $J_r(\lambda)$ as above. In this case $x_s = \lambda 1_r$ and $x_u = \lambda^{-1} x$ clearly satisfy the conditions of (2.7), and hence x_s and x_u are defined for all $x \in GL_n(K)$: they are called respectively the *semisimple part* and the *unipotent part* of x.

Now let G be any linear algebraic group, and let $x \in G$. By (2.5) there exists an injective homomorphism of algebraic groups $\varphi : G \to GL_n(K)$ for some n. In this situation it can be shown that the semisimple and unipotent parts of the matrix $\varphi(x)$ lie in $\varphi(G)$, and more precisely that the elements x_s, $x_u \in G$ defined by $\varphi(x_s) = \varphi(x)_s$ and $\varphi(x_u) = \varphi(x)_u$ depend only on x and not on the embedding φ of G in a general linear group. As in the previous case we have

$$x = x_s x_u = x_u x_s$$

and x_s, x_u are called respectively the *semisimple part* and the *unipotent part* of $x \in G$. Moreover,

(2.8) *Let $\varphi : G \to H$ be a homomorphism of linear algebraic groups, and*

2 Linear algebraic groups: definition and elementary properties 153

let $x \in G$. Then

$$\varphi(x)_s = \varphi(x_s), \quad \varphi(x)_u = \varphi(x_u).$$

An element $x \in G$ is *semisimple* if $x = x_s$ (i.e., if $x_u = e$, the identity element of G), and $x \in G$ is *unipotent* if $x = x_u$ (i.e., if $x_s = e$). Let G_s (resp. G_u) denote the subset of G consisting of semisimple (resp. unipotent) elements.

(2.9) (i) G_u is closed in G.
(ii) G_s is a constructible subset of G.
(iii) $G_s \cap G_u = \{e\}$.

Proof. We may assume that G is a closed subgroup of $GL_n(K)$. Hence it is enough to prove (2.9) when $G = GL_n(K)$. Now $x \in GL_n(K)$ is unipotent if and only if $(x - 1_n)^n = 0$, which shows that the set of unipotent matrices is closed in $GL_n(K)$, proving (i). Next, when $G = GL_n(K)$ we have $x \in G_s$ if and only if $x \in gD_ng^{-1}$ for some $g \in G$, where D_n is the diagonal subgroup of G; hence G_s is the image of the morphism $\varphi : G \times D_n \to G$ defined by $\varphi(g, t) = gtg^{-1}$, hence is constructible by (1.12). Finally, if $x \in G_s \cap G_u$ then $x_u = x_s = e$ and therefore $x = e$.

In general, G_s and G_u are *not* subgroups of G. However,

(2.10) *Let G be a commutative linear algebraic group. Then G_s and G_u are closed subgroups of G, and $\mu : G_s \times G_u \to G$ is an isomorphism of algebraic groups.*

Proof. Since G is commutative it follows that G_s and G_u are subgroups of G; G_u is closed by (2.9)(i), and G_s is closed by (2.9)(ii) and (2.2)(ii). So certainly $\mu : G_s \times G_u \to G$ is a bijective homomorphism of linear algebraic groups. On the other hand, the mapping $x \mapsto x_s$ is a morphism, hence so is $\mu^{-1} : x \mapsto (x_s, xx_s^{-1})$.

Interlude

A linear algebraic group G is said to be *unipotent* if $G = G_u$, i.e, if each $x \in G$ is unipotent. For example, the group U_n (§2, Ex. 4(d)) is unipotent, and so are all its closed subgroups. Conversely, in fact, every unipotent group is isomorphic to a closed subgroup of some U_n.

Next, G is said to be *solvable* if it is solvable as an abstract group, that is to say if the "derived series" $(D^n G)_{n \geqslant 0}$ reaches $\{e\}$ in a finite number of steps, where $D^0 G = G$ and (for $n \geqslant 0$) $D^{n+1} G = (D^n G, D^n G)$ is the group generated by all commutators $(x, y) = xyx^{-1}y^{-1}$ with $x, y \in D^n G$. In particular, unipotent groups are solvable; on the other hand, the group B_n (§2, Ex. 4(c)) of upper triangular $n \times n$ matrices is solvable but not unipotent.

Now let G be any linear algebraic group. The *radical* $R(G)$ (resp. *unipotent radical* $R_u(G)$) of G is the unique maximal, closed, connected, solvable (resp. unipotent), normal subgroup of G. We have $R_u(G) \subset R(G)$, and indeed $R_u(G) = R(G)_u$.

If $R_u(G) = \{e\}$, the group G is said to be *reductive*. If $R(G) = \{e\}$, it is said to be *semisimple*.

We now have the following chain of subgroups in an arbitrary linear algebraic group G:

```
                          G
          (finite)        |
                          G_0       (connected)
       (semisimple)       |
                         R(G)       (solvable)
          (torus)         |
                         R_u(G)     (unipotent)
       (unipotent)        |
                         {e}
```

2 Interlude

where (as in §2) G_0 is the identity component of G. The entries on the left of the chain describe the successive quotients: thus (as we have already seen in §2) G/G_0 is a finite group, $G_0/R(G)$ is semisimple, and $R(G)/R_u(G)$ is an (algebraic) *torus*, isomorphic to a product of copies of the multiplicative group $\mathbb{G}_m$.

One of the aims of these lectures (although I shall have to skip a lot of the details) will be indicate how a *connected reductive* linear algebraic group (i.e. the quotient $G_0/R_u(G)$ in the chain above) is classified up to isomorphism by a combinatorial object called its *root datum*, which is a slightly more elaborate version of the root systems of Roger Carter's lectures. This classification is independent of the underlying (algebraically closed) field K.

At the same time this classifies the *compact connected Lie groups*. For if U is a compact connected Lie group then the $\mathbb{C}$-algebra $C_{\text{alg}}(U)$ spanned by the matrix coefficients of the finite-dimensional representations of U is the affine algebra $\mathbb{C}[G]$ of a connected reductive linear algebraic group G over $\mathbb{C}$ (the *complexification* of U, cf. [Segal, 3.9]). In the other direction, U is (isomorphic to) a maximal compact subgroup of G. So we have a one-one correspondence between (isomorphism classes of) compact connected Lie groups U and (isomorphism classes of) connected reductive linear algebraic groups G over the field of complex numbers. In terms of Lie algebras, if $\mathfrak{u}$ (resp. $\mathfrak{g}$) is the Lie algebra of U (resp. G), then $\mathfrak{g}$ is the complexification of $\mathfrak{u}$, and $\mathfrak{u}$ is the compact real form of $\mathfrak{g}$. The group G is semisimple if and only if U has finite centre, and G is an (algebraic) torus if and only if U is a (geometric) torus, i.e. a product of copies of the circle group $\{z \in \mathbb{C} : |z| = 1\}$.

In talking of the successive quotients in the chain of subgroups of G above, I have run ahead of myself because I have not yet shown how to factor out a normal closed subgroup H of a linear algebraic group G. Let $A = K[G]$ be the affine algebra of G; then H acts on A as follows:

$$h\varphi(x) = \varphi(xh)$$

for $h \in H$, $x \in G$ and $\varphi \in A$. Hence

$$A^H = \{\varphi \in A : h\varphi = \varphi, \text{ all } h \in H\}$$

consists of the functions $\varphi \in A$ constant on each coset of H in G. In fact A^H is the affine algebra of a linear algebraic group G/H, and the embedding $A^H \hookrightarrow A$ is dual to a surjective homomorphism of G onto G/H with kernel H.

However, we shall also need to consider G/H when H is a closed (but not normal) subgroup of G. (The analogous situation in Lie theory is that of a Lie group G and a closed subgroup H, and one shows that the set $G/H = \{xH : x \in G\}$ has a natural structure of a smooth manifold.) In the algebraic context, we shall show that G/H can be given the structure of an algebraic variety, which in general is *not* affine. Thus the little algebraic geometry covered in §1 is no longer adequate; we need to introduce projective and quasi-projective varieties.

3
Projective algebraic varieties

If G is a linear algebraic group and H is a closed subgroup of G, we shall see in §6 that the set $X = G/H$ of cosets xH ($x \in G$) can be endowed with the structure of an algebraic variety. However, X is not always an *affine* variety: in general it is a quasi-projective variety (to be defined below). Thus we need to develop a more general notion of algebraic variety.

Let X be an affine algebraic variety, let U be a non-empty open subset of X, and let $x \in U$. A function f on U with values in K is said to be *regular at* x if there exists an open neighbourhood U' of x contained in U, and functions $g, h \in K[X]$ such that h vanishes nowhere on U' and $f(y) = g(y)h(y)^{-1}$ for all $y \in U'$. Then it is a basic fact (which requires proof) that

(3.1) *A function $f : X \to K$ is regular if and only if f is regular at each $x \in X$.*

For each non-empty open set $U \subset X$ let $\mathcal{O}(U) = \mathcal{O}_x(U)$ denote the K-algebra of functions $f : U \to K$ that are regular at each $x \in U$. Then:

(a) *If $V \subset U$ are non-empty open sets in X and $f \in \mathcal{O}(U)$ then $f|V \in \mathcal{O}(V)$.*
(b) *If a non-empty open set $U \subset X$ is covered by non-empty open subsets U_i, and if $f : U \to K$ is such that $f|U_i \in \mathcal{O}(U_i)$ for each i, then $f \in \mathcal{O}(U)$.*

These two conditions (a) and (b) say that the assignment $U \mapsto \mathcal{O}_X(U)$ (for U open in X) is a *sheaf* $\mathcal{O}_X$ of functions on X, called the *structure sheaf* of the affine variety X. From this point of view, the pair $(X, \mathcal{O}_X)$ is a *ringed space*, i.e. a topological space carrying a sheaf of functions (satisfying the conditions (a) and (b) above). From (3.1) it follows that the affine algebra $K[X]$ of X is just $\mathcal{O}_X(X)$, and it may appear to the reader that all we have achieved by this discussion is to replace a simple object $K[X]$ by a more complicated object, namely the sheaf $\mathcal{O}_X$. But this elaboration has a purpose, namely to enable us to define a more general notion of algebraic varieties.

Prevarieties and varieties

Let $(X, \mathcal{O})$ be a ringed space and let Y be a subset of X. Give Y the induced topology, and for each non-empty open subset V of Y let $(\mathcal{O}|Y)(V)$ be the set of functions $f : V \to K$ such that for each $x \in V$ there is an open neighbourhood U_x of x in X and a function $f_x \in \mathcal{O}(U_x)$ which agrees with f on $V \cap U_x$. The assignment $V \mapsto (\mathcal{O}|Y)(V)$ is a *sheaf* $\mathcal{O}|Y$ on Y (i.e., it satisfies conditions (a) and (b) above), called the sheaf *induced* by $\mathcal{O}$ on Y. In particular, if Y is open in X we have $(\mathcal{O}|Y)(V) = \mathcal{O}(V)$ for all V open in Y.

Next let $(X, \mathcal{O}_X)$ and $(Y, \mathcal{O}_Y)$ be ringed spaces, and let $\varphi : X \to Y$ be a continuous map. Then φ is a *morphism of ringed spaces* if, for each open $V \subset Y$ and each $f \in \mathcal{O}_Y(V)$, the function $f \circ \varphi : \varphi^{-1}V \to K$ belongs to $\mathcal{O}_X(\varphi^{-1}V)$. When X and Y are affine algebraic varieties, this notion of morphism agrees with that defined in §1.

After these preliminaries, a *prevariety* (over K) is a ringed space $(X, \mathcal{O}_X)$ such that X is covered by a finite number of open sets U_i with the property that each induced ringed space $(U_i, \mathcal{O}|U_i)$ is isomorphic (as ringed space) to an affine algebraic variety. Intuitively, X is obtained by patching together a finite number of affine varieties in such a way that the regular functions agree on the overlaps.

It follows from this definition that X is a Noetherian topological space, so that (1.5) applies to X.

We shall habitually drop $\mathcal{O}_X$ from the notation, and speak of X (rather than $(X, \mathcal{O}_X)$) as a prevariety; and the reader may be relieved to be told that (except in the present section) he will never see the structure sheaf $\mathcal{O}_X$ explicitly referred to; but he should bear in mind that it is always implicitly present, as an essential part of the structure of X.

One shows next that if X and Y are prevarieties, the product $X \times Y$ (satisfying the usual universal property) exists and is unique up to isomorphism. Briefly, if $X = \bigcup_{i=1}^{m} U_i$ and $Y = \bigcup_{j=1}^{n} V_j$, where the U_i and V_j are affine open sets, then $X \times Y$ is covered by the products $U_i \times V_j$, which are themselves affine varieties.

3 Projective algebraic varieties

A *variety* is a prevariety X satisfying the separation property (1.10); the diagonal $\Delta_X = \{(x,x) : x \in X\}$ is closed in $X \times X$.

If X is an irreducible variety covered by affine open sets U_i, each U_i is irreducible by (1.4) (since the closure of U_i is X) and each intersection $U_i \cap U_j$ is non-empty. It follows that U_i and U_j have the same function field, which is called the *function field* $K(X)$ of X. The *dimension* of X is defined, as in the affine case, to be the transcendence degree of $K(X)$ over K.

Finally, propositions (1.12) — (1.14) remain true for arbitrary varieties.

Projective Varieties

The most important examples of non-affine varieties (and the only ones we shall encounter) are the projective and quasi-projective varieties, to which we now turn.

If V is a finite-dimensional K-vector space, the *projective space* $P(V)$ of V is the set of all lines (i.e. 1-dimensional subspaces) in V. If $V = K^{n+1}$, $P(V)$ is denoted by $P_n(K)$. A line in K^{n+1} is determined by any point $(x_0, \ldots, x_n) \neq 0$ on it, and hence a point $x \in P_n(K)$ has $n+1$ *homogeneous coordinates* $(x_0, \ldots, x_n)$, not all zero, and such that $(x_0, \ldots, x_n)$ and $(\lambda x_0, \ldots, \lambda x_n)$, where λ is any non-zero element of K, represent the same point of $P_n(K)$.

In projective geometry, equations of varieties are homogeneous; so if $S \subset K[t_0, \ldots, t_n]$ is any set of homogeneous polynomials (not necessarily of the same degree), let

$$V(S) = \{x \in P_n(K) : f(x) = 0, \text{ all } f \in S\}.$$

Just as in the affine case, the $V(S)$ are the closed sets in a topology (the Zariski topology) on $P_n(K)$.

Consider in particular $H_i = V(t_i)$, a hyperplane in P_n whose complement U_i is the open set consisting of all $x = (x_0, \ldots, x_n) \in P_n$ such that $x_i \neq 0$. By homogeneity we may assume that $x_i = 1$, so that U_i consists of all $x \in P_n$ with coordinates $(x_0, \ldots, x_{i-1}, 1, x_{i+1}, \ldots, x_n)$ and hence is in bijective correspondence with affine space K^n. Thus we have

$$P_n(K) = U_0 \cup U_1 \cup \ldots \cup U_n$$

the union of $n+1$ open sets each identified with K^n. We have a structure sheaf $\mathcal{O}_{U_i}$ on each U_i, and since the restrictions of $\mathcal{O}_{U_i}$ and $\mathcal{O}_{U_j}$ to $U_i \cap U_j$ coincide, the sheaves $\mathcal{O}_{U_i}$ are the restrictions to U_i of a well-defined sheaf $\mathcal{O} = \mathcal{O}_{P_n}$ on P_n, the *structure sheaf* of P_n. Thus projective space $P_n(K)$ is a prevariety as defined above, and it may be verified that it is in fact a *variety* (the diagonal of $P_n \times P_n$ is a closed set).

A *projective* (resp. *quasi-projective*) *variety* X is now defined to be a closed (resp. locally closed) subset of a projective space $P_n(K)$, together with its induced structure sheaf $\mathcal{O}_X = \mathcal{O}_{P_n}|X$. Both affine and projective varieties are quasi-projective.

On an affine variety X, as we saw in §1, there are plenty of regular functions defined on all of X; indeed enough to determine the structure of X. On a projective variety, on the other hand, this is not the case: if X is an irreducible projective variety, we have $\mathcal{O}_X(X) = K$, i.e. the only regular functions defined on all of X are the constant functions. (Compare Liouville's theorem: the only holomorphic functions on the Riemann sphere ($= P_1(\mathbb{C})$) are the constants.)

Complete varieties

An algebraic variety X is said to be *complete* if for any variety Y the projection morphism $X \times Y \to Y$ is a closed mapping, i.e. maps closed sets to closed sets. (This notion is an analogue for algebraic varieties of the notion of compactness in the category of locally compact (Hausdorff) topological spaces: if X is locally compact, then X is compact if and only if, for all locally compact spaces Y, the projection $X \times Y \to Y$ is a closed mapping.)

(3.2) *Let X, Y be varieties.*

(i) *If X is complete and Y is closed in X, then Y is complete.*

(ii) *If X and Y are complete, then $X \times Y$ is complete.*

(iii) *If $\varphi : X \to Y$ is a morphism and X is complete, then $\varphi(X)$ is closed in Y and is complete.*

(iv) *If Y is a complete subvariety of X, then Y is closed in X.*

(v) *If X is complete and irreducible, the only regular functions on X are the constant functions.*

(vi) *If X is affine and complete then X is finite (i.e. $\dim X = 0$).*

Proof. (i) and (ii) are immediate from the definitions. As to (iii), let $\Gamma = \{(x, \varphi(x)) : x \in X\} \subset X \times Y$ be the graph of φ. Then Γ is the inverse image of the diagonal Δ_Y under the morphism $(x, y) \mapsto (\varphi(x), y)$ of $X \times Y$ into $Y \times Y$, hence is closed in $X \times Y$ (because Y is a variety). Since $\varphi(X)$ is the image of Γ under the projection $X \times Y \to Y$, it follows that $\varphi(X)$ is closed in Y. To show that $\varphi(X)$ is complete, we

3 Projective algebraic varieties

may assume that $\varphi(X) = Y$. If Z is any variety, let $p : X \times Z \to Z$, $q : Y \times Z \to Z$ be the projections; then if W is closed in $Y \times Z$, we have $q(W) = p((\varphi \times 1)^{-1}(W))$ closed in Z, because X is complete. (iv) now follows from (iii), applied to the inclusion morphism $Y \to X$. Next, a regular function f on X may be regarded as a morphism of X into the affine line K^1, which is also a morphism of X into the projective line P_1. If X is irreducible, then $f(X)$ is an irreducible proper closed subset of P_1, by (1.4) and (iii) above, hence consists of a single point (because the only proper closed subsets of P_1 are the finite subsets). Hence f is constant, which proves (v). Finally, (vi) is a direct consequence of (v).

A basic fact (which we shall not prove here) is that

(3.3) *Projective varieties are complete.*

In view of (3.2)(i), it is enough to show that projective space $P = P_n(K)$ is complete, and for this it is enough to show that the projection $P \times K^m \to K^m$ is a closed mapping, for each $m \geq 0$.

In view of (3.3), the assertions of (3.2) apply to projective varieties. In particular, the image of a projective variety X under a morphism $\varphi : X \to Y$ is a closed subset of Y (contrast with the analogous statement (1.12) for affine varieties).

4
Tangent spaces. Separability

Let X be an affine algebraic variety, embedded as a closed subset in K^n, and let $f_1, \ldots, f_r \in K[t_1, \ldots, t_n]$ be a set of generators of the ideal $I(X)$, so that for $x \in K^n$ we have $x \in X$ if and only if $f_i(x) = 0$ ($1 \leq i \leq r$).

Suppose for the moment that $K = \mathbb{C}$. To say that a vector $v \in \mathbb{C}^n$ is a tangent vector to X at $x \in X$ means that $f_i(x + \epsilon v)$ is $O(\epsilon^2)$ for small $\epsilon \in \mathbb{C}$ and $i = 1, 2, \ldots, r$, or equivalently that

(*) $$f_i(x + \epsilon v) \equiv 0 \pmod{\epsilon^2} \qquad (1 \leq i \leq r)$$

as polynomials in ϵ.

This condition makes sense for any field K, and we may reformulate it as follows. The algebra $D = D(K)$ of *dual numbers* over K is defined to be

$$D = K[t]/(t^2) = K \oplus K\epsilon$$

where ϵ is the image of t in D, so that $\epsilon^2 = 0$. Thus the elements of D are of the form $a + b\epsilon$ with $a, b \in K$, and add and multiply as follows:

$$(a + b\epsilon) + (a' + b'\epsilon) = (a + a') + (b + b')\epsilon,$$
$$(a + b\epsilon)(a' + b'\epsilon) = aa' + (ab' + a'b)\epsilon.$$

Hence $a + b\epsilon \mapsto a$ is a K-algebra homomorphism of D onto K.

The condition (*) for a tangent vector $v \in K^n$ is now replaced by

(4.1) $$f_i(x + \epsilon v) = 0 \qquad (1 \leq i \leq r)$$

and we are therefore led to the following definition: *a vector $v \in K^n$ is a tangent vector to X at x if and only if the mapping $f \mapsto f(x + \epsilon v)$ of $K[X]$ into D is a K-algebra homomorphism.*

Accordingly we define the *tangent bundle* of X to be

$$T(X) = \mathrm{Hom}_{K-\mathrm{alg}}(K[X], D)$$

and the projection $a + b\epsilon \mapsto a$ of D onto K projects $T(X)$ onto $X =$ $\text{Hom}_{K-\text{alg}}(K[X], K)$. For each $x \in X$, the fibre $T_x(X)$ of $T(X)$ over x is the K-vector space of tangent vectors at x.

If $\xi \in T_x(X)$ and $f \in K[X]$, then $\xi(f)$ is of the form

(4.2) $$\xi(f) = f(x) + \epsilon d_\xi(f)$$

where $d_\xi : K[X] \to K$ satisfies

(4.3) $$d_\xi(fg) = d_\xi(f)g(x) + f(x)d_\xi(g)$$

for all $f, g \in K[X]$, since $\xi(fg) = \xi(f)\xi(g)$. Conversely, each K-linear mapping $d_\xi : K[X] \to K$ satisfying (4.3) determines a tangent vector $\xi \in T_x(X)$ by the formula (4.2). One should think of $d_\xi f$ as the derivative of f at x in the direction ξ.

Now let $\varphi : X \to Y$ be a morphism of affine algebraic varieties, so that we have a K-algebra homomorphism $\varphi^* : K[Y] \to K[X]$, and hence

$$T(\varphi) : T(X) \to T(Y)$$

namely $\xi \mapsto \xi \circ \varphi^*$ for $\xi \in T(X)$. Over each $x \in X$ the restriction of $T(\varphi)$ to $T_x(X)$ is a K-linear mapping

$$T_x(\varphi) : T_x(X) \to T_{\varphi(x)}(Y).$$

If $\xi \in T_x(X)$ and $\eta = T_x(\varphi)\xi$, then

(4.4) $$d_\eta = d_\xi \circ \varphi^*.$$

If $\psi : Y \to Z$ is another morphism of affine algebraic varieties we have $T(\psi \circ \varphi) = T(\psi) \circ T(\varphi)$, so that

(4.5) $$T_x(\psi \circ \varphi) = T_{\varphi(x)}(\psi) \circ T_x(\varphi)$$

which is just the "chain rule" of differential calculus.

Finally, if X is any algebraic variety as in §3, covered by affine open sets U_i, one checks that the tangent bundles $T(U_i)$ patch together to form the *tangent bundle* $T(X)$ of X, and that $T(X)$ depends only on X and not on the affine open covering chosen. For each $x \in X$, the fibre $T_x(X)$ of $T(X)$ over x is a finite-dimensional K-vector space.

(4.6) *Example.* Let V be a finite-dimensional K-vector space, $P = P(V)$ the projective space of V (§3), and let $\pi : V - \{0\} \to P$ be the projection. Let $x \in V - \{0\}$, so that $\pi(x)$ is the line $L = Kx$ generated by x. The tangent space $T_x(V)$ may be identified with V; with this identification $T_x(\pi) : V \to T_L(P)$ is surjective with kernel L. (We may take $V = K^{n+1}$

and $x = (x_0, x_1, \ldots, x_n)$ with $x_0 \neq 0$, so that $\pi(x) \in U_0$ in the notation of §3, and $\pi(x) = (x_0^{-1} x_1, \ldots, x_0^{-1} x_n)$. Let $\xi = (\xi_0, \ldots, \xi_n) \in T_x(V)$. Since

$$\frac{x_i + \epsilon \xi_i}{x_0 + \epsilon \xi_0} = \frac{1}{x_0}(x_i + \epsilon \xi_i)(1 - \frac{\epsilon \xi_0}{x_0})$$

$$= \frac{x_i}{x_0} + \frac{\epsilon}{x_0^2}(x_0 \xi_i - x_i \xi_0)$$

it follows that $T_x(\pi)\xi = \eta = (\eta_1, \ldots, \eta_n)$, where $\eta_i = x_0^{-2}(x_0 \xi_i - x_i \xi_0)$, and hence that $T_x(\pi)\xi = 0$ if and only if $x_0 \xi_i = x_i \xi_0$ ($1 \leq i \leq n$), that is to say if and only if $\xi \in Kx$.)

If X is an arbitrary variety, in general the tangent spaces $T_x(X)$ will not all have the same dimension, even if X is irreducible. For example, if X is the plane curve with equation $x_1 x_2 + x_1^3 + x_2^3 = 0$, which has a double point at the origin, then $T_x(X)$ is 1-dimensional for all $x \in X$ except $(0,0)$, and $T_{(0,0)}(X) = K^2$ is 2-dimensional. In fact we have

(4.7) *Let X be an irreducible variety of dimension d. Then*
(i) $\dim T_x(X) \geq d$ *for all $x \in X$.*
(ii) $S = \{x \in X : \dim T_x(X) > d\}$ *is a proper closed subvariety of X.*

S is called the *singular locus* of X. If $\dim T_x(X) = d$, we say that x is a *simple point* of X, or that X is *smooth* at x. If every $x \in X$ is simple (i.e. if S is empty) we say that X is *smooth*.

Now let G be a connected linear algebraic group. For each $x \in G$, let $\lambda_x : G \to G$ be left multiplication by x, as in §2. Then $T_e(\lambda_x) : T_e(G) \to T_x(G)$ is an isomorphism, so that $\dim T_x(G)$ is the same at all $x \in G$, hence by (4.7) is equal to $\dim G$, and G is *smooth*.

Separability

At this point it is instructive to return to Example (1.8). $\varphi : \mathbb{G}_a \to \mathbb{G}_a$ is the mapping defined by $\varphi(x) = x^p$, where $p > 0$ is the characteristic of K. This mapping φ is a homomorphism because $(x + y)^p = x^p + y^p$, hence is a bijective homomorphism of algebraic groups, but is *not* an automorphism of the algebraic group $\mathbb{G}_a$. Let $\xi \in T_x(\mathbb{G}_a) = K$: since $(x + \epsilon \xi)^p = x^p + \epsilon^p \xi^p = x^p$, it follows that $T(\varphi)$ maps each tangent space $T_x(\mathbb{G}_a)$ to 0; thus φ has an 'infinitesimal' kernel $T_0(\mathbb{G}_a)$, invisible to the naked eye.

The notion of *separability*, which we shall now introduce briefly, is designed to avoid this sort of situation. If $E \subset F$ are fields such that F is finitely generated over E, then F is said to be *separably generated* over E

4 Tangent spaces. Separability

if there is an intermediate field E' such that E'/E is a pure transcendental extension and F/E' is a finite separable (algebraic) extension (so that $F = E'(\alpha)$ where the minimal polynomial for α over E' has no repeated roots).

Now suppose that X and Y are irreducible varieties, with respective function fields, $K(X)$, $K(Y)$. A *dominant* morphism $\varphi : X \to Y$ (i.e., such that $\varphi(X)$ is dense in Y) induces an embedding φ^* of $K(Y)$ in $K(X)$, and φ is said to be *separable* if $K(X)$ is separably generated over $\varphi^*K(Y)$.

A more geometrical criterion for separability is contained in the following proposition:

(4.8) *Let $\varphi : X \to Y$ be a morphism between irreducible varieties.*
(i) *Suppose that $x \in X$ and $y = \varphi(x) \in Y$ are simple points of X, Y respectively and that $T_x(\varphi) : T_x(X) \to T_y(Y)$ is surjective. Then φ is dominant and separable.*
(ii) *Conversely, suppose that φ is dominant and separable. Then there is a non-empty open subset U of X such that for each $x \in U$ the point $y = \varphi(x)$ is simple on Y and $T_x(\varphi) : T_x(X) \to T_y(Y)$ is surjective.*

Remark. In characteristic zero all finitely generated field extensions are separably generated, and hence all dominant morphisms are separable: inseparability is a "characteristic p" phenomenon.

5
The Lie algebra of a linear algebraic group

Let G be a linear algebraic group, $A = K[G]$ its affine algebra, and let $\mathfrak{g} = T_e(G)$ be the tangent space to G at the identity element e. From §4 we know that $\mathfrak{g}$ is a K-vector space of dimension equal to that of G.

We shall show that $\mathfrak{g}$ can be given the structure of a Lie algebra over K, by interpreting the elements of $\mathfrak{g}$ as derivations. More precisely, let $L(G)$ denote the space of all K-linear maps $\delta : A \to A$ which (a) are derivations, i.e. satisfy

(5.1) $$\delta(fg) = \delta(f)g + f\delta(g)$$

for all $f, g \in A$; and (b) are *left-invariant*, i.e. satisfy

(5.2) $$\lambda(x)\delta = \delta\lambda(x)$$

for all $x \in G$, where as in §2 $\lambda(x)f$ is the function $y \mapsto f(x^{-1}y)$. If $\delta_1, \delta_2 \in L(G)$ one checks immediately that $\delta = [\delta_1, \delta_2] = \delta_1\delta_2 - \delta_2\delta_1$ satisfies (5.1) and (5.2), and hence that $L(G)$ is a Lie algebra over K.

Recall from §4 that each $X \in \mathfrak{g}$ determines a K-linear mapping

$$d_X : A \to K$$

by the rule $X(f) = f(e) + d_X(f)\epsilon$, so that

(5.3) $$d_X(fg) = d_X(f)g(e) + f(e)d_X(g)$$

for $f, g \in A$. Now define, for each $X \in \mathfrak{g}$ and $f \in A$, a function $\delta_X f$ on G by the rule

(5.4) $$(\delta_X f)(x) = d_X(\lambda(x^{-1})f).$$

Let us first check that $\delta_X f \in A$. As in §2, let $\mu : G \times G \to G$ be the product map and let

$$\mu^* f = \sum g_i \otimes h_i \qquad (g_i, h_i \in A)$$

5 The Lie algebra of a linear algebraic group

so that $f(xy) = \sum g_i(x)h_i(y)$ for all $x, y \in G$, and therefore

$$\lambda(x^{-1})f = \sum g_i(x)h_i,$$

from which it follows that

$$\delta_X f = \sum g_i d_X(h_i) \in A.$$

Thus an equivalent definition of δ_X is

$$\delta_X = (1 \otimes d_X) \circ \mu^*.$$

(5.5) *The mapping* $X \mapsto \delta_X$ *is a K-linear isomorphism of* $\mathfrak{g}$ *onto* $L(G)$.

Proof. From (5.3) and (5.4) it follows that δ_X is a K-derivation of A which is left-invariant, because for all $f \in A$ and $x, y \in G$ we have

$$\begin{aligned}(\lambda(y)\delta_X f)(x) &= (\delta_X f)(y^{-1}x) = d_X(\lambda((y^{-1}x)^{-1})f) \\ &= d_X(\lambda(x^{-1})\lambda(y)f) = \delta_X(\lambda(y)f)(x)\end{aligned}$$

so that $\lambda(y) \circ \delta_X = \delta_X \circ \lambda(y)$.

Conversely, if $\delta \in L(G)$ the mapping $d : A \to K$ defined by $df = (\delta f)(e)$ satisfies

$$d(fg) = d(f)g(e) + f(e)d(g)$$

and hence (§4) $d = d_X$ for a unique $X \in \mathfrak{g}$; and by left-invariance we have

$$\begin{aligned}(\delta f)(x) &= (\lambda(x^{-1})\delta f)(e) = \delta(\lambda(x^{-1})f)(e) \\ &= d_X(\lambda(x^{-1})f) = (\delta_X f)(x)\end{aligned}$$

so that $\delta = \delta_X$, completing the proof.

In view of (5.5) we may transport the Lie algebra structure of $L(G)$ to $\mathfrak{g}$ by defining

(5.6) $$\delta_{[X,Y]} = [\delta_X, \delta_Y] = \delta_X\delta_Y - \delta_Y\delta_X$$

for all $X, Y \in \mathfrak{g}$. $L(G)$ (or $\mathfrak{g}$) is the *Lie algebra* of G.

(5.7) *Example.* Let $G = GL_n(K)$, so that $A = K[G] = K[t_{ij}(1 \leq i, j \leq n), d^{-1}]$, where the t_{ij} are the coordinate functions on G, and $d = \det(t_{ij})$. Since G is an open subset of matrix space $M_n(K) = K^{n^2}$, the tangent vectors at the identity $1_n \in G$ are the homomorphisms $f \mapsto f(1_n + \epsilon X)$, where $X = (X_{ij}) \in M_n(K)$, and we may identify $\mathfrak{g} = T_{1_n}(G)$ with $M_n(K)$. We have $d_X(t_{ij}) = X_{ij}$ and hence, for $x = (x_{ij}) \in G$,

$$\begin{aligned}(\delta_X t_{ij})(x) &= d_X(\lambda(x^{-1})t_{ij}) \\ &= d_X(\sum_{k=1}^n x_{ik} t_{kj}) \\ &= \sum_{k=1}^n x_{ik} X_{kj}\end{aligned}$$

so that $\delta_X T = TX$, where T is the matrix (t_{ij}). From (5.6) it follows that the Lie algebra structure on $\mathfrak{g} = M_n(K)$ is the usual one: $[X, Y] = XY - YX$. This Lie algebra is denoted by $\mathfrak{gl}_n(K)$.

More generally, if V is any finite-dimensional K-vector space, the Lie algebra $\mathfrak{gl}(V)$ of the algebraic group $GL(V)$ is the Lie algebra of the associative algebra $\mathrm{End}_K(V)$ of all K-linear maps $V \to V$.

Let now $\varphi : G \to H$ be a homomorphism of linear algebraic groups, and let $\mathfrak{g}, \mathfrak{h}$ be the Lie algebras of G, H respectively. The *differential* of φ is

$$d\varphi = T_e(\varphi) : \mathfrak{g} \to \mathfrak{h}.$$

From the definitions it follows that if $X \in \mathfrak{g}$ and $Y = (d\varphi)X \in \mathfrak{h}$, then

(5.8) (i) $d_Y = d_X \circ \varphi^*$,
(ii) $\varphi^* \circ \delta_Y = \delta_X \circ \varphi^*$,

from which it is easy to check that $d\varphi : \mathfrak{g} \to \mathfrak{h}$ is a Lie algebra homomorphism.

Let V be a finite-dimensional subspace of $A = K[G]$, stable under right translation $\rho(x)$ for each $x \in G$, so that we have a homomorphism of algebraic groups $\rho : G \to GL(V)$. Let $X \in \mathfrak{g}$ and let $Y = (d\rho)X \in \mathfrak{gl}(V)$. Then we have

(5.9) $\delta_X f = Yf$ for all $f \in V$.

Proof. Choose a K-basis $f_1, \ldots, f_n$ of V, thereby identifying $GL(V)$ with $GL_n(K)$. As in §2 we have

(1) $$\mu^* f_j = \sum_{i=1}^n f_i \otimes \varphi_{ij}$$

5 The Lie algebra of a linear algebraic group

for suitable $\varphi_{ij} \in A$, so that

$$\rho(x)f_j = \sum_i \varphi_{ij}(x)f_i$$

and $\rho(x)$ is identified with the matrix $(\varphi_{ij}(x)) \in GL_n(K)$.

Also from (1) we have

$$\lambda(x^{-1})f_j = \sum_i f_i(x)\varphi_{ij}$$

so that

(2) $$\delta_X f_j = \sum_i d_X(\varphi_{ij})f_i.$$

But $d_X\varphi_{ij} = d_X\varphi^*(t_{ij}) = d_Y t_{ij} = Y_{ij}$ by (5.7) and (5.8), hence (2) becomes

$$\delta_X f_j = \sum_i Y_{ij}f_i = Yf_j.$$

Let G be a linear algebraic group, H a closed subgroup of G, and let $\mathfrak{a} = \{f \in A : f|H = 0\}$ be the ideal in $A = K[G]$ defined by H. Restriction of functions to H defines a K-algebra isomorphism $A/\mathfrak{a} \xrightarrow{\sim} K[H]$, and hence the tangent bundle of H may be identified with the set of $\xi \in T(G)$ such that $\xi(\mathfrak{a}) = 0$. In particular, the Lie algebra $\mathfrak{h} = T_e(H)$ of H is identified with the vector subspace of $\mathfrak{g}$ consisting of the $X \in \mathfrak{g}$ such that $d_X\mathfrak{a} = 0$.

(5.10) (i) $H = \{x \in G : \rho(x)\mathfrak{a} \subset \mathfrak{a}\} = \{x \in G : \lambda(x)\mathfrak{a} \subset \mathfrak{a}\}$.
(ii) $\mathfrak{h} = \{X \in \mathfrak{g} : \delta_X(\mathfrak{a}) \subset \mathfrak{a}\}$.

Proof. (i) Let $x \in H$, $f \in \mathfrak{a}$. For all $y \in H$ we have $(\rho(x)f)(y) = f(yx) = 0$, whence $\rho(x)f \in \mathfrak{a}$ and therefore $\rho(x)\mathfrak{a} \subset \mathfrak{a}$. Conversely, if $\rho(x)\mathfrak{a} \subset \mathfrak{a}$ and $f \in \mathfrak{a}$, then $f(x) = (\rho(x)f)(e) = 0$, so that $x \in H$. Likewise with ρ replaced by λ.
(ii) Let $X \in \mathfrak{g}$. Then

$$\begin{aligned}
X \in \mathfrak{h} &\Leftrightarrow d_X\mathfrak{a} = 0 \\
&\Leftrightarrow d_X(\lambda(x^{-1})\mathfrak{a}) = 0 \quad \text{for all} \quad x \in H \quad \text{(by (i))} \\
&\Leftrightarrow (\delta_X f)(x) = 0 \quad \text{for all} \quad f \in \mathfrak{a},\ x \in H \\
&\Leftrightarrow \delta_X\mathfrak{a} \subset \mathfrak{a}.
\end{aligned}$$

(5.11) *Let $\varphi : G \to H$ be a bijective homomorphism of linear algebraic groups, and let $\mathfrak{g}$, $\mathfrak{h}$ be the Lie algebras of G, H respectively. Then φ is an isomorphism if and only if $d\varphi : \mathfrak{g} \to \mathfrak{h}$ is an isomorphism of Lie algebras.*

Clearly, if φ is an isomorphism so is $d\varphi$. In the other direction we shall merely indicate the proof. We may assume that G, and therefore also H, is connected. If $E = K(G)$ and $F = K(H)$ are the function fields, it can be shown that the field extension E/φ^*F is purely inseparable. If now $d\varphi$ is an isomorphism, it follows from (4.3)(i) that E/φ^*F is separable; hence $E = \varphi^*F$, so that φ^* and hence also φ is an isomorphism.

In particular, if K has characteristic 0, every bijective homomorphism is an isomorphism (because there are no inseparable field extensions).

The adjoint representation

Let G be a linear algebraic group, $\mathfrak{g}$ its Lie algebra. For each $x \in G$ we have an *inner automorphism* $\text{Int}\, x : G \to G$, namely $g \mapsto xgx^{-1}$. Its differential is denoted by $\text{Ad}\, x$:

$$\text{Ad}\, x = d(\text{Int}\, x) : \mathfrak{g} \to \mathfrak{g}.$$

This is a linear transformation of the vector space $\mathfrak{g}$, with inverse

$$(\text{Ad}\, x)^{-1} = \text{Ad}(x^{-1}),$$

so that $\text{Ad}\, x \in GL(\mathfrak{g})$. Moreover, since $\text{Int}(xy) = (\text{Int}\, x) \circ (\text{Int}\, y)$ for all x, $y \in G$, it follows that $\text{Ad}(xy) = (\text{Ad}\, x) \circ (\text{Ad}\, y)$ and hence that

$$\text{Ad} = \text{Ad}_G : G \to GL(\mathfrak{g})$$

is a *representation* of G, called the *adjoint representation*.

(5.12) *Let* $\mathsf{X} \in \mathfrak{g}$, $\mathsf{Y} = (\text{Ad}\, x)\mathsf{X}$. *Then*

$$\delta_\mathsf{Y} = \rho(x) \circ \delta_\mathsf{X} \circ \rho(x^{-1}).$$

Proof. Let $\varphi = \text{Int}\, x$, so that $(\varphi^*f)(y) = f(xyx^{-1})$ for $f \in A = K[G]$ and $y \in G$, and therefore $\varphi^* = \lambda(x^{-1})\rho(x^{-1})$. The formula (5.12) now follows from (5.8)(ii), being in mind that δ_X commutes with $\lambda(x^{-1})$.

(5.13) *Example.* Let $G = GL_n(K)$, $x \in G$, $\mathsf{X} \in \mathfrak{g}$. Then

$$(\text{Ad}\, x)\mathsf{X} = x\mathsf{X}x^{-1}.$$

Proof. Let $T = (t_{ij})$ be the matrix of coordinate functions on G, as in (5.7). If $\mathsf{Y} = (\text{Ad}\, x)\mathsf{X}$ we have from (5.7) and (5.12)

$$\begin{aligned} T\mathsf{Y} = \delta_\mathsf{Y} T &= \rho(x)\delta_\mathsf{X}\rho(x^{-1})T = \rho(x)\delta_\mathsf{X} Tx^{-1} \\ &= \rho(x)T\mathsf{X}x^{-1} = Tx\mathsf{X}x^{-1}. \end{aligned}$$

5 The Lie algebra of a linear algebraic group

From (5.13) it follows that $\mathrm{Ad}_G : G \to GL(\mathfrak{g})$ is a homomorphism of algebraic groups in the particular case that $G = GL_n(K)$. If now G is any linear algebraic group, we may by (2.5) embed G is a closed subgroup in some $GL_n(K)$, and correspondingly $\mathfrak{g}$ as a subspace of $\mathfrak{gl}_n(K)$. One checks that Ad_G is the restriction to G of $\mathrm{Ad}_{GL_n(K)}$, and hence

(5.14) *For any linear algebraic group G, $\mathrm{Ad} : G \to GL_n(\mathfrak{g})$ is a homomorphism of algebraic groups.*

6
Homogeneous spaces and quotients

Let G be a linear algebraic group, X an algebraic variety. An *action* of G on X is a morphism $G \times X \to X$, written $(g, x) \mapsto gx$, such that

(i) $g(hx) = (gh)x$ for all $g, h \in G$ and $x \in X$,
(ii) $ex = x$ for all $x \in X$.

The variety X, equipped with an action of G as above, is called a *G-variety* or *G-space*. If X and Y are G-spaces, a morphism $\varphi : X \to Y$ is a *G-morphism* if $\varphi(gx) = g\varphi(x)$ for all $g \in G$ and $x \in X$.

Let X be a G-space, $x \in X$. The *isotropy group* of x in G is the group $G_x = \{g \in G : gx = x\}$, the subgroup of G that fixes x. It is a closed subgroup of G, because it is the inverse image of $\{x\}$ under the morphism $g \mapsto gx$ of G into X. The *orbit* of x is $Gx = \{gx : g \in G\}$, which is the image of G under the same morphism $g \mapsto gx$, hence by (1.12) contains a non-empty open subset U of $\overline{Gx}$. Since $\overline{Gx}$ is G-stable, it follows that the translates gU of U cover Gx, and hence that Gx is open in its closure $\overline{Gx}$. This proves the first part of

(6.1) (i) *Each orbit of G in X is a locally closed subvariety of X.*
(ii) *If G is connected, all orbits of minimal dimension are closed. In particular, there is at least one closed orbit.*

As to (ii), let Y be an orbit of minimum dimension d. (Since G is connected, Y is irreducible.) Now G acts on $\overline{Y}$, hence $Z = \overline{Y} - Y$ is a union of orbits of G, all of dimension $< d$, since $\dim Z < d$. We conclude that Z is empty, i.e. that Y is closed.

A *homogeneous space* for G is a G-space X on which G acts transitively (i.e., there is only one orbit). In that case the isotropy groups G_x, $x \in X$, are all conjugate in G. If we fix a point $x_0 \in X$ and write $H = G_{x_0}$ (a closed subgroup of G) and G/H for the set of cosets gH ($g \in G$), the mapping $gx_0 \mapsto gH : X \to G/H$ is bijective.

Let H be a closed subgroup of G. A *quotient* of G by H is a pair (X, x_0) consisting of a homogeneous space X for G together with a point $x_0 \in X$ with isotropy group H, such that the following universal property holds: for any pair (Y, y_0) consisting of a homogeneous space Y for G and a point $y_0 \in Y$ whose isotropy group contains H, there is a unique G-morphism $\varphi : X \to Y$ such that $\varphi(x_0) = y_0$. Clearly, if a quotient exists, it is unique up to G-isomorphism.

When is a homogeneous space for G a quotient? One answer to this question is provided by (6.2) below, but we shall have to omit the proof. Suppose X is a homogeneous space for G, and that $x_0 \in X$ has isotropy group $G_{x_0} = H$. Let $\psi : G \to X$ be the mapping $g \mapsto gx_0$, and let $X_0 = \psi(G_0)$, where G_0 is the identity component of G. Let $\psi_0 : G_0 \to X_0$ be the restriction of ψ to G_0.

(6.2) *(X, x_0) is a quotient of G by H if and only if the morphism $\psi_0 : G_0 \to X_0$ is separable.*

Consider the morphism ψ. Its fibres are the cosets gH of H in G, hence are all of dimension $\dim H$, and X is a variety of dimension $\dim G - \dim H$. Moreover, by homogeneity all the tangent spaces $T_x(X)$ ($x \in X$) have the same dimension, so that X is smooth and $\dim T_x(X) = \dim G - \dim H$ by (4.2).

Let $\mathfrak{g}$, $\mathfrak{h}$ be the Lie algebras of G, H respectively. Then we can reformulate (6.2) as follows:

(6.3) *Let X be a homogeneous space for G, and let $x_0 \in X$ have isotropy group H. Then (X, x_0) is a quotient of G by H if and only if the kernel of $T_e(\psi) : \mathfrak{g} \to T_{x_0}(X)$ is equal to $\mathfrak{h}$.*

Proof. We may assume that G is connected. Since ψ maps H to the point x_0, the kernel of $T_e(\psi)$ contains $\mathfrak{h}$. Since the dimension of $T_{x_0}(X)$ is $\dim G - \dim H = \dim \mathfrak{g} - \dim \mathfrak{h}$, it follows that $\ker T_e(\psi) = \mathfrak{h}$ if and only if $T_e(\psi)$ is surjective, that is to say (4.3) if and only if ψ is separable. Hence (6.3) follows from (6.2).

Now let G be any linear algebraic group and H a closed subgroup of G. Let $\mathfrak{g}$, $\mathfrak{h}$ be the Lie algebras of G, H respectively, so that $\mathfrak{h}$ is a subalgebra of $\mathfrak{g}$. We shall show how to construct a quotient of G by H.

For this purpose let $A = K[G]$ and let $\mathfrak{a} = \{f \in A : f|H = 0\}$ be the ideal defined by H. Since A is a Noetherian ring, the ideal $\mathfrak{a}$ is finitely generated, say by $f_1,\ldots,f_r$. By (2.4) there is a finite-dimensional vector subspace V of A, stable under G (acting by right translations) and containing $f_1,\ldots,f_r$. Let $\rho : G \to GL(V)$ be the representation of G so defined, and let $U = V \cap \mathfrak{a}$, which is a subspace of V containing the generators f_i of the ideal $\mathfrak{a}$. Then we have

(6.4) (i) $H = \{x \in G : \rho(x)U = U\}$,
(ii) $\mathfrak{h} = \{X \in \mathfrak{g} : d\rho(X)U \subset U\}$.

Proof. (i) Let $x \in H$, $f \in U$. Then $(\rho(x)f)(y) = f(yx)$ vanishes for all $y \in H$ (since $f \in \mathfrak{a}$), hence $\rho(x)f \in V \cap \mathfrak{a} = U$ and therefore $\rho(x)U \subset U$. But $\rho(x)$ is bijective, hence $\rho(x)U = U$. Conversely, if $\rho(x)U = U$ then $\rho(x)f_i \in \mathfrak{a}$ for each generator f_i, and therefore $f_i(x) = (\rho(x)f_i)(e) = 0$, whence $x \in H$.

(ii) By (5.9) we have $d\rho(X)U = \delta_X U$ for all $X \in \mathfrak{g}$. If $X \in \mathfrak{h}$ then $\delta_X U \subset V \cap \mathfrak{a} = U$ by (5.10). Conversely, if $\delta_X U \subset U$ then $\delta_X f_i \in \mathfrak{a}$ for each i. Since the f_i generate $\mathfrak{a}$ and δ_X is a derivation, it follows that $\delta_X \mathfrak{a} \subset \mathfrak{a}$ and hence that $X \in \mathfrak{h}$ by (5.10) again.

The next step is to squash U down to a line (i.e. a one-dimensional space). To do this we take exterior powers: if U has dimension d then $L = \wedge^d U$ is a line contained in $E = \wedge^d V$. The group G acts on E via $\wedge^d \rho = \varphi$, say:

$$\varphi(x)(v_1 \wedge \ldots \wedge v_d) = \rho(x)v_1 \wedge \ldots \wedge \rho(x)v_d,$$
$$d\varphi(X)(v_1 \wedge \ldots \wedge v_d) = \sum_{i=1}^{d} v_1 \wedge \ldots \wedge \delta_X v_i \wedge \ldots \wedge v_d$$

for $x \in G$, $X \in \mathfrak{g}$ and $v_1,\ldots,v_d \in V$. From these formulas and (6.4) it is not difficult to deduce that

(6.5) (i) $H = \{x \in G : \rho(x)L = L\}$,
(ii) $\mathfrak{h} = \{X \in \mathfrak{g} : d\varphi(X)L \subset L\}$.

We now pass to the projective space $P = P(E)$, in which L is a point, say $x_0 \in P$. The group G acts on P via φ, and by (6.5)(i) the isotropy group G_{x_0} is precisely H. Let $X = Gx_0$ be the orbit of $x_0 (= L)$ in P. By (6.1), X is locally closed in P, that is to say it is a quasi-projective variety on which G acts transitively. In fact (X, x_0) *is a quotient of G by H*.

To show this, it is enough by (6.3) to verify that the kernel of $T_e(\psi)$ is equal to $\mathfrak{h}$. Choose a non-zero element $u_0 \in L$ and let $\omega : G \to E$ be the

6 Homogeneous spaces and quotients

mapping $g \mapsto \varphi(g)u_0$; also let $\pi : E - \{0\} \to P$ be the projection, so that $\pi(u) = Ku$ for $u \in E$, $u \neq 0$. Then $\psi = \pi \circ \omega$ and hence

$$T_e(\psi)X = T_{u_0}(\pi)(d\varphi(X)u_0)$$

for $X \in \mathfrak{g}$. Now by (4.6) the kernel of $T_{u_0}(\pi)$ is $L = Ku_0$, hence by (6.5)(ii) the kernel of $T_e(\psi)$ is $\mathfrak{h}$, as required.

We have therefore proved

(6.6) *If G is any linear algebraic group, H a closed subgroup of G, then a quotient of G by H exists and is a quasi-projective variety.*

We denote the quotient of G by H by G/H.

(6.7) Example. Let $G = GL_n(K)$, acting on the space $V = K^n$ of column vectors of length n, with basis consisting of the unit vectors $e_1, \ldots, e_n$. A (complete) *flag* in V is a sequence of subspaces

$$\mathfrak{f} : 0 = U_0 < U_1 < \ldots < U_n = V$$

such that $\dim U_i = i$ for each i. In particular,

$$\mathfrak{f}_0 : 0 = V_0 < V_1 < \ldots < V_n = V$$

is a flag, where V_i is spanned by $e_1, \ldots, e_i$. If $g \in G$, then $g\mathfrak{f}$ is the flag

$$0 = U_0 < gU_1 < \ldots < gU_{n-1} < U_n = V.$$

Let F denote the set of all flags in V. Then G acts transitively on F: if $\mathfrak{f}$ is as above, choose vectors $u_1, \ldots, u_n$ such that $u_i \in U_i - U_{i-1}$ ($1 \leqslant i \leqslant n$); then $u_1, \ldots, u_i$ is a basis of U_i for each i, and $\mathfrak{f} = g\mathfrak{f}_0$ where g is the element of G defined by $ge_i = u_i$ ($1 \leqslant i \leqslant n$). Moreover the subgroup B of G that fixes $\mathfrak{f}_0$ is the group of upper triangular matrices $g = (g_{ij})$, $g_{ij} = 0$ if $i > j$.

The set F may be given the structure of a projective algebraic variety, as follows. With $u_1, \ldots, u_n$ as above, the exterior product $u_1 \wedge \ldots \wedge u_i$ depends (up to a nonzero scalar multiple) only on U_i, and hence the tensor product

$$(*) \quad u_1 \otimes (u_1 \wedge u_2) \otimes (u_1 \wedge u_2 \wedge u_3) \otimes \ldots \in V \otimes \wedge^2 V \otimes \wedge^3 V \otimes \ldots = E,$$

say, depends (up to a non-zero scalar multiple) only on the flag $\mathfrak{f}$. Hence if $P(E)$ is the projective space of E, we have a mapping $\varphi : F \to P(E)$, namely $\mathfrak{f} \mapsto$ image of $(*)$ in $P(E)$. It can be verified that φ is injective and that $X = \varphi(F)$ is a closed subvariety of $P(E)$, hence a projective variety.

Moreover if $x_0 \in X$ is the image of the flag $\mathfrak{f}_0$, then (X, x_0) is a quotient of G by B. Thus G/B is a projective variety, called the *flag variety*.

Finally, when H is normal in G, we can say more:

(6.8) *Let G be a linear algebraic group, H a closed normal subgroup of G. Then G/H is an affine variety and, when provided with the usual group structure, a linear algebraic group.*

7
Borel subgroups and maximal tori

The key to unravelling the structure of a compact connected Lie group G is the theorem that the maximal tori of G are all conjugate in G. In the case of a linear algebraic group, the key to unravelling its structure is the study of its connected closed *solvable* subgroups, and for this purpose the following fixed-point theorem is crucial:

(7.1) *Let G be a connected solvable linear algebraic group acting on a nonempty projective variety X. Then G has a fixed point in X.*

Proof (sketch). We proceed by induction on $d = \dim G$. If $d = 0$ then $G = \{e\}$, and the result is obvious. So assume $d > 0$ and let $G' = DG$ be the derived group of G, generated by all commutators $xyx^{-1}y^{-1}$ ($x, y \in G$). In fact G' is closed and connected and (of course) solvable, and is a proper subgroup of G, so that $\dim G' < d$. Hence, by the inductive hypothesis, G' has a fixed point in X. For each $g \in G'$ the set $X^g = \{x \in X : gx = x\}$ is closed in X, because it is the inverse image of the diagonal $\Delta = \{(x, x) : x \in X\}$ under the morphism $x \mapsto (x, gx)$ of X into $X \times X$, and Δ is closed in X since X is projective. Hence

$$X' = \bigcap_{g \in G'} X^g,$$

the set of fixed points of G', is closed (and non-empty), hence projective. Moreover, G' is normal in G and hence X' is G-stable.

By (6.1), there exists $x \in X'$ such that the orbit Gx is closed, hence projective. On the other hand, the isotropy group G_x of x contains G', hence is normal in G (because G/G' is an abelian group). To complete the proof we need the following result, whose proof we omit:

(7.2) *Let Y, Z be irreducible transitive G-spaces, and let $\varphi : Y \to Z$ be a bijective G-morphism. If Z is complete, so is Y.*

We take $Y = G/G_x$ and $Z = Gx$, φ being the morphism $gG_x \mapsto gx$. Z is projective, hence complete, and therefore Y is complete. But also Y is affine, since it is the quotient of G by a normal subgroup; and irreducible, since G is connected. Hence by (3.2)(vi) Y is a single point, so that $G_x = G$ and x is a fixed point.

An immediate corollary of (7.1) is

(7.3) *Let G be a connected closed solvable subgroup of $GL_n(K)$. Then there exists $g \in G$ such that $G \subset gBg^{-1}$, where $B = B_n$ is the group of upper triangular matrices.*

Proof. G acts on the flag variety F of K^n (6.7), which is a projective variety. By (7.1) there is a flag $\mathfrak{f}$ fixed by G. If $\mathfrak{f} = g\mathfrak{f}_0$ in the notation of (6.7), we have $G \subset gBg^{-1}$.

Borel subgroups

Let G be a connected linear algebraic group. Consider the set of all connected closed solvable subgroups of G, ordered by inclusion. This set has maximal elements, for reasons of dimension. These maximal elements are the *Borel subgroups* of G. For example, if $G = GL_n(K)$ it follows from (7.3) that the Borel subgroups are precisely the conjugates of B_n, and therefore are all conjugate in G. We shall now show that this is so in general.

(7.4) *Let G be a connected linear algebraic group and let B be a Borel subgroup of G. Then*
(i) *G/B is a projective variety.*
(ii) *All Borel subgroups of G are conjugate to B.*

Proof. Choose a Borel subgroup B_0 of largest possible dimension. By (6.5) there exists a K-vector space of dimension (say) n, a line V_1 in V and a linear action of G on V such that B_0 is the stabiliser of V_1. Thus B_0 acts on V/V_1, and by (7.3) there is a flag in V/V_1, say $0 < V_2/V_1 < \cdots < V/V_1$, stabilised by B_0. It follows that the stabiliser of the flag
$$\mathfrak{f} : 0 < V_1 < V_2 < \cdots < V_n = V$$
is precisely B_0, and hence we have a bijective morphism of G/B_0 onto the orbit $G\mathfrak{f}$.

If $\mathfrak{f}'$ is any other flag in V, its isotropy group is solvable, hence has

7 Borel subgroups and maximal tori

dimension $\leqslant \dim B_0$, and its orbit $G\mathfrak{f}'$ has dimension $\geqslant \dim G - \dim B_0 = \dim G\mathfrak{f}$. Hence $G\mathfrak{f}$ is an orbit of minimum dimension in the flag variety $F(V)$, hence by (6.1) is closed in $F(V)$ and therefore projective. By (7.2) it follows that G/B_0 is complete; since it is also quasi-projective (6.11) it follows that G/B_0 is a projective variety.

Now let B be any Borel subgroup of G. It acts on G/B_0 by left multiplication. By the fixed point theorem (7.1) there exists $x \in G$ such that $BxB_0 = xB_0$, whence $B \subset xB_0x^{-1}$. But xB_0x^{-1} is a Borel subgroup (since B_0 is), hence $B = xB_0x^{-1}$.

(7.5) *Let P be a closed subgroup of G. Then G/P is a projective variety if and only if P contains a Borel subgroup.*

Proof. Suppose G/P is projective, and let B be a Borel subgroup of G. Then B acts on G/P by left multiplication, and by (7.1) this action has a fixed point, i.e. there exists $x \in G$ such that $BxP = xP$ and hence $Bx \subset xP$, i.e. $P \supset x^{-1}Bx$.

Conversely, if $P \supset B$ there is a surjective morphism $G/B \to G/P$. By (3.2)(iii), G/P is complete and therefore projective.

Subgroups P of G satisfying the equivalent conditions of (7.5) are called *parabolic* subgroups of G.

(7.6) *Example.* Let $G = GL_n(K)$, acting on $V = K^n$ as usual, and let $v = (n_1, \ldots, n_r)$ be any sequence of positive integers such that $n_1 + \ldots + n_r = n$. A *flag of type v* in V is a sequence

$$\mathfrak{f} : 0 = U_0 < U_1 < \ldots < U_r = V$$

of subspaces of V such that $\dim U_i - \dim U_{i-1} = n_i$ ($1 \leqslant i \leqslant r$). The group G acts transitively on the set $F_v(V)$ of flags of type v, and the stabiliser of any flag of type v is a parabolic subgroup P_v of G. If $v = (1, \ldots, 1)$, P_v is a Borel subgroup, at the other extreme, if $v = (n)$, then $P_v = G$. There are 2^{n-1} choices for v, and correspondingly 2^{n-1} conjugacy classes of parabolic subgroups in $GL_n(K)$.

Maximal tori

From here on, for lack of space and time, our account will become increasingly sketchy. An (algebraic) *torus* is a linear algebraic group isomorphic to $\mathbb{G}_m^n = \mathbb{G}_m \times \ldots \times \mathbb{G}_m$ (n factors) for some $n \geqslant 1$; or, equivalently, to the group $D_n(K)$ of diagonal matrices. (Warning: in the theory of Lie groups, a torus is a compact abelian group isomorphic to a product of copies of the circle S^1: not the same thing at all.)

A torus is connected and abelian, hence solvable, and all its elements are semisimple. Conversely (but not obviously) if G is a connected abelian linear algebraic group all of whose elements are semisimple, then G is a torus.

Let now G be any linear algebraic group. Just as in the case of solvable subgroups, the set of closed tori in G, ordered by inclusion, has maximal elements, for reasons of dimension. These are the *maximal tori* of G. We want to show that they are all conjugate in G. Since they all lie in the identity component G_0 of G, we may as well assume that G is *connected*.

Each maximal torus is connected and solvable, and hence is contained in some Borel subgroup B of G. Hence it is enough to show that all maximal tori in B are conjugate in B, i.e. we reduce to the situation where G is *solvable* (and connected).

We need now to consider the structure of connected solvable groups.

(7.7) *Example.* Let $G = B_n$, the group of $n \times n$ upper triangular matrices. This is the semidirect product $D_n \ltimes U_n$, where D_n is the group of diagonal matrices, and U_n consists of the upper triangular matrices with 1's down the diagonal. Thus $U_n = G_u$, the set of unipotent elements of G, and therefore G_u is a connected closed normal subgroup of G. Moreover, D_n is a maximal torus in G. For if T is a closed torus in G, then $T \cap U_n = \{1_n\}$, because each element of T is semisimple and each element of U_n is unipotent, hence (since U_n is normal) TU_n is a constructible subgroup of G, hence closed, and $T = TU_n/U_n$ has dimension $\leq \dim G - \dim U_n = \dim D_n$.

This example is typical. The following theorem (which I shall not prove here) elucidates the structure of connected solvable group:

(7.8) *Let G be a connected solvable linear algebraic group. Then*

(i) *G_u is a connected closed normal subgroup of G.*

(ii) *The maximal tori in G are all conjugate, and if T is one then G is the semidirect product of T and G_u.*

(By (7.3) we may assume that G is a closed subgroup of B_n, whence $G_u = G \cap U_n$, which is a closed normal subgroup of G since by (7.7) U_n is a closed normal subgroup of B_n. But it is less obvious that G_u is connected, and the proof of (ii) is quite complicated: there is apparently no nice fixed-point theorem as in the case of Borel subgroups ((7.4) above).)

From (7.8) and the conjugacy of Borel subgroups (7.4) it follows immediately that

(7.9) *Let G be any linear algebraic group. Then*
(i) *The maximal tori in G are all conjugate.*
(ii) *The maximal connected unipotent subgroups of G are all conjugate.*

(In (ii), a maximal connected unipotent subgroup U of G is contained in a Borel subgroup B, hence is equal to B_u.)

8
The root structure of a linear algebraic group

Characters and one-parameter subgroups of tori

A *character* of a torus T is a homomorphism of algebraic groups

$$\chi : T \to \mathbb{G}_m.$$

If $t \in T$ we shall write t^χ rather than $\chi(t)$ for the image of t under χ. If χ, ψ are characters of T then so are $\chi + \psi$ and $-\chi$, defined by $t^{\chi+\psi} = t^\chi t^\psi$ and $t^{-\chi} = (t^\chi)^{-1}$. Hence the set $X(T)$ of characters of T is in a natural way an abelian group.

Dually, a *one-parameter subgroup* of T (by abuse of language) is a homomorphism of algebraic groups

$$\eta : \mathbb{G}_m \to T.$$

As above, we shall write x^η in place of $\eta(x)$ ($x \in \mathbb{G}_m$). The set $Y(T)$ of one-parameter subgroups of T is again an abelian group: $x^{\eta+\zeta} = x^\eta . x^\zeta$, $x^{-\eta} = (x^\eta)^{-1}$.

The structure of these groups $X(T)$, $Y(T)$ is easily described. Consider first the case $T = \mathbb{G}_m$. It is easily verified that the characters of $\mathbb{G}_m$ are just the mappings $x \mapsto x^r$, $r \in \mathbb{Z}$. If now T is a torus of dimension d, say $T = \mathbb{G}_m^d$, then $X(T) \cong X(\mathbb{G}_m)^d \cong \mathbb{Z}^d$ and $Y(T) \cong Y(\mathbb{G}_m)^d \cong \mathbb{Z}^d$. More precisely, for each $v = (n_1, \ldots, n_d) \in \mathbb{Z}^d$ define $\chi_v \in X(T)$ and $\eta_v \in Y(T)$ by $\chi_v(x_1, \ldots, x_d) = x_1^{n_1} \ldots x_d^{n_d}$ and $\eta_v(x) = (x^{n_1}, \ldots, x^{n_d})$. Then $v \mapsto \chi_v$ and $v \mapsto \eta_v$ are isomorphisms of $\mathbb{Z}^d$ onto $X(T)$ and $Y(T)$ respectively.

Finally we have a pairing

$$X(T) \times Y(T) \to \mathbb{Z},$$

say $(\chi, \eta) \mapsto \langle \chi, \eta \rangle$, defined as follows:

$$(x^\eta)^\chi = x^{\langle \chi, \eta \rangle}.$$

8 The root structure of a linear algebraic group

This pairing is linear in each variable, and puts the two lattices (=free abelian groups) $X(T), Y(T)$ in duality. In the notation introduced above if $\mu = (m_1, \ldots, m_d)$ and $\nu = (n_1, \ldots, n_d)$ we have $\langle \chi_\mu, \eta_\nu \rangle = \sum_1^d m_i n_i$.

The root system $R(G, T)$.

Let G be a linear algebraic group, T a maximal torus in G. Recall from §5 the adjoint representation $\mathrm{Ad}_G : G \to GL(\mathfrak{g})$, where $\mathfrak{g}$ is the Lie algebra of G. Since Ad_G is a homomorphism of algebraic groups, $\mathrm{Ad}_G(T)$ consists of commuting semisimple elements, hence is diagonalizable: this means that, relative to the action of T, $\mathfrak{g}$ decomposes as a direct sum

$$\mathfrak{g} = \bigoplus_{\alpha \in X(T)} \mathfrak{g}_\alpha$$

where for each character $\alpha \in X(T)$,

$$\mathfrak{g}_\alpha = \{ X \in \mathfrak{g} : \mathrm{Ad}_G(t)X = \alpha(t)X, \text{ all } t \in T \}.$$

The nonzero $\alpha \in X(T)$ such that $\mathfrak{g}_\alpha \neq 0$ are called the *roots* of G relative to T, and the set of roots is denoted by $R(G, T)$: it is a finite subset of $X(T)$.

Let $Z_G(T)$ and $N_G(T)$ denote respectively the centralizer and normalizer of T in G. Then in fact $Z_G(T)$ is the identity component of $N_G(T)$ (this follows from the so-called "rigidity" of tori: a torus has no connected set of automorphisms consisting of more than one element). Hence

$$W = W(G, T) = N_G(T)/Z_G(T)$$

is a finite group, which acts on T, hence also on $X(T)$, and permutes the roots $\alpha \in R(G, T)$. Specifically, if $w \in W$ is the image of $n \in N_G(T)$ then $(w\alpha)(t) = \alpha(n^{-1}tn)$ for $t \in T$. W is the *Weyl group* of $R(G, T)$.

The root datum $\mathscr{R}(G, T)$

The notion of a *root datum* is a fancier version of that of a root system, used for example to classify semisimple Lie algebras. Roughly speaking, it consists of a root system R embedded in a lattice X (but R need not span X) together with the dual root system $R^\vee$ (obtained by reversing the arrows in the Dynkin diagram of R) embedded in the dual lattice $Y = \mathrm{Hom}(X, \mathbb{Z})$. The formal definition runs as follows. A *root datum* $\mathscr{R} = (X, Y, R, R^\vee)$ consists of

(1) Lattices (i.e., free abelian groups) X and Y and a bilinear mapping $X \times Y \to \mathbb{Z}$, written $(\xi, \eta) \mapsto \langle \xi, \eta \rangle$, inducing isomorphisms $X \mapsto \text{Hom}(Y, \mathbb{Z})$ and $Y \mapsto \text{Hom}(X, \mathbb{Z})$.

(2) finite subsets $R \subset X$ and $R^\vee \subset Y$, and a bijection $\alpha \mapsto \alpha^\vee$ of R onto $R^\vee$.

For each $\alpha \in R$ define $s_\alpha : X \to X$ by

$$s_\alpha \xi = \xi - \langle \xi, \alpha^\vee \rangle \alpha$$

and $s_{\alpha^\vee} : Y \to Y$ by

$$s_{\alpha^\vee} \eta = \eta - \langle \alpha, \eta \rangle \alpha^\vee.$$

Then the axioms are

(RD1) $\quad \langle \alpha, \alpha^\vee \rangle = 2,$
(RD2) $\quad s_\alpha R \subset R, \quad s_{\alpha^\vee} R^\vee \subset R^\vee$

for all $\alpha \in R$.

Let Q be the subgroup of X generated by R and let $V = Q \otimes \mathbb{R}$. If $R \neq \emptyset$ then R is a root system in V in the usual sense, and $R^\vee$ is the dual root system (in the dual vector space V^*).

The root datum $\mathscr{R}$ is *reduced* if

(RD3) $\quad \alpha \in R \Rightarrow 2\alpha \notin R.$

Starting with a linear algebraic group G and a maximal torus T in G (which, as we have seen, is unique up to conjugacy) we have already constructed three of the four ingredients $X, Y, R, R^\vee$: X is $X(T)$, Y is $Y(T)$ and R is $R(G, T)$ as defined above. It remains to construct $R^\vee$, and one way of doing this is as follows. Let $\alpha \in R$ and let $T_\alpha = (\text{Ker}\,\alpha)_0$, the connected component of the identity in the kernel of α; then T_α is a subtorus of T of codimension 1. Let G_α be the centralizer of T_α in G; then it turns out that G_α is connected, and clearly T is a maximal torus of G_α. The Weyl group $W(G_\alpha, T)$ in fact has order 2, and embeds in $W(G, T)$. Let $s_\alpha \in W(G, T)$ be the non-identity element of $W(G_\alpha, T)$; then s_α acts on $X(T)$ as follows:

$$s_\alpha(\chi) = \chi - \langle \chi, \alpha^\vee \rangle \alpha$$

for a unique $\alpha^\vee \in Y(T)$. Since $s_\alpha^2 = 1$ we have $\langle \alpha, \alpha^\vee \rangle = 2$. Then $R^\vee = R^\vee(G, T)$ is by definition the set of these $\alpha^\vee$, for all $\alpha \in R$.

In this way we can associate with any linear algebraic group G and a maximal torus T in G a root datum $\mathscr{R}(G, T)$, which depends up to

8 The root structure of a linear algebraic group

isomorphism only on G, since different choices of T are conjugate in G, and hence may be denoted simply by $\mathscr{R}(G)$. It can be shown that $\mathscr{R}(G)$ is *reduced*, i.e. satisfies (RD3).

Example. Let $G = GL_n(K)$, $T = D_n(K)$. Then we may identify $X(T)$ and $Y(T)$ with $\mathbb{Z}^n$, the pairing being the usual scalar product on $\mathbb{Z}^n$. The roots $\alpha \in R(G, T)$ are the characters $t \mapsto t_i t_j^{-1}$ ($i \neq j$), where $t = \text{diag}(t_1, \ldots, t_n)$ (because $(\text{Ad}\, t)x = txt^{-1}$ for $x \in G$). Hence if $\epsilon_1, \ldots, \epsilon_n$ is the standard basis of $\mathbb{Z}^n$, we have $R(G, T) = \{\epsilon_i - \epsilon_j : i \neq j\}$. If we follow through the construction of $R^\vee \subset Y(T)$ given above, we find that with α as above we have $\alpha^\vee = \epsilon_i - \epsilon_j$.

We recall that a linear algebraic group G is said to be *reductive* if $R_u(G) = \{e\}$, that is to say if G contains no connected closed unipotent normal subgroup other than $\{e\}$. We shall conclude with the statement of the following *existence and uniqueness theorem* for connected reductive groups over an algebraically closed field K:

(8.1) (Uniqueness) *Let G_1, G_2 be connected reductive linear algebraic groups over K. Then G_1, G_2 are isomorphic as algebraic groups if and only if $\mathscr{R}(G_1) \cong \mathscr{R}(G_2)$.*

(Existence) *Let $\mathscr{R}$ be a reduced root datum. Then there exists a connected reductive linear algebraic group G over K such that $\mathscr{R}(G) \cong \mathscr{R}$.*

Notes and references

§1. The background algebraic geometry is developed *ab initio* in [S], Chapter 1 and (modulo some commutative algebra) in [H], Chapter I and [B], Chapter AG. These references contain proofs of all the propositions of §1.

§2. See [B], Chapter I; [H], Chapter II and Chapter VI (for the Jordan decomposition); [S] Chapter 2.

§3. See [B], Chapter AG; [H], Chapter I; [S] Chapter 1.

§4. See [B], Chapter AG §§16, 17; [H], Chapter I, §5; [S], Chapter 3.

§5. See [B], Chapter I, §3; [H], Chapter III; [S], Chapter 3.

§6. See [B], Chapter II; [H], Chapter IV; [S], Chapter 5.

§7. See [B], Chapter IV: §11; [H], Chapter VIII, §21; [S], Chapters 6 and 7.

§8. The proof of the existence and uniqueness theorem is given in full in [S], Chapter 11 and 12.

Bibliography

J.F. Adams. *Lectures on Lie groups*. Benjamin, 1969. Also avaliable as Midway reprint, University of Chicago, 1982.

A. Borel, *Linear Algebraic Groups*. Math. Lecture Note Series, W.A. Benjamin, Inc. New York, 1969.

N. Bourbaki, *Groups et algebras de Lie IV, V, VI*, Masson, Paris, 1981.

E. Cartan, *Oeuvres Complètes*, Gauthier-Villars, Paris, 1952.

R.W. Carter, Simple groups and simple Lie algebras, *J. London Math. Soc.* **40**, 193–240, 1965.

R.W. Carter, *Finite groups of Lie type, conjugacy classes and complex characters*, Wiley Classics Library Edition, J. Wiley, New York, 1989.

R.W. Carter, *Simple Groups of Lie Type*. Wiley Classics Library Edition, J. Wiley, New York, 1989.

C. Chevalley, *Theory of Lie groups*. Princeton University Press, 1946.

C. Chevalley, Sur certains groups simples, *Tôhuku Math. J.* **7**, 14–66, 1955.

I.M. Gelfand, M.I. Graev, N.Ya. Vilenkin, *Generalized functions*. Academic Press, 1966.

P. Griffiths and J. Harris, *Principles of algebraic geometry*. Wiley, 1978.

S. Helgason, *Differential geometry, Lie groups, and symmetric spaces*. Academic Press, 1978.

G. Hochschild, *The structure of Lie groups*. Holden-Day, 1965.

J.E. Humphreys, *Introduction to Lie Algebras and Representation Theory*. Graduate Texts in Mathematics, Springer-Verlag, New York, 1972.

J.E. Humphreys, *Linear Algebraic Groups*. Graduate Texts in Mathematics, Springer-Verlag, New York, 1975.

N. Jacobson, *Lie Algebras*. Interscience Publishers, J. Wiley, New York, 1962.

W. Killing, Die Zusammensetzung der stetigen endlichen Transformationsgruppen I–IV, *Math. Ann.* **31**, 252–290, 1988; **33**, 1–48, 1889; **34**, 57–122, 1889; **36**, 161–189, 1890.

A. Kirillov, *Elements of the theory of representations*. Springer-Verlag, Berlin, 1976.

A.W. Knapp, *Representation theory of semisimple groups*. Princeton University Press, 1986.

J. Milnor [1], *Morse theory*. Ann. of Math. Studies **51**, Princeton University Press, 1963.

J. Milnor [2], *Introduction to algebraic K-theory*. Ann. of Math. Studies **72**, Princeton University Press, 1971.

J. Milnor [3], *Remarks on infinite-dimensional Lie groups*. In *Proc. of Summer School on Quantum Gravity*, ed. B. DeWitt, Les Houches, 1983.

J. Milnor and J. Stasheff, *Characteristic classes*. Ann. of Math. Studies **76**, Princeton University Press, 1974.

D. Montgomery and L. Zippin, *Topological transformation groups*. Interscience, 1955.

A. Pressley and G. Segal, *Loop Groups*. Oxford University Press, 1986.

R. Ree, A family of simple groups associated with the simple Lie algebra of type (F_4), *Amer. J. Math.* **83**, 401–420, 1961.

R. Ree, A family of simple groups associated with the simple Lie algebra of type (G_2), *Amer. J. Math.* **83**, 432–462, 1961.

J.-P. Serre, *Lie algebras and Lie groups*. Benjamin, 1965.

T.A. Springer, *Linear Algebraic Groups*. Birkhäuser, Boston, 1981

R. Steinberg, Variations on a theme of Chevalley, *Pacific J. Math.* **9**, 875–891, 1959.

M. Suzuki, On a class of doubly transitive groups, *Ann. of Math.* **75**, 105–145, 1962.

J. Tits, Algèbra alternatives, algèbras de Jordan et alègbras de Lie exceptionnelles, *Indag. Math.* **28**, 223–237, 1966.

Index

additive group, 147
adjoint representation, 170
affine algebra, 142
affine algebraic variety, 142
algebraic set, 139
algebraic torus, 155
anti-self-dual, 55
atlas, 69

Borel subgroups, 178
Borel-Weil theorem, 116
braid group, 113
Bruhat decomposition, 66, 119

Campbell-Baker-Hausdorff series, 75
Cartan decomposition, 14
Cartan matrix, 17
Cartan subalgebra, 12
Cartan subgroups, 127
Cayley parametrization, 70
celestial sphere, 56
characters, 95, 182
charts, 69
Chevalley basis, 37
Chevalley group, 38
classification of finite dimensional
  irreducible g-modules, 28
classification of finite dimensional simple
  Lie algebras, 22
classification of the finite simple groups, 44
Clifford algebra, 132
complementary series, 127
complete, 160
complex structures, 61, 119
connected, 148
constructible, 145
coordinate ring, 142

differential, 168
dimension, 145

Dirac's spanner, 76
discrete series, 125
dominant, 143
dominant weight, 117
dual numbers, 162
Dynkin diagram, 18

Euclidean geometry, 49
exponential map, 73

flag, 175
flag of type v, 179
flag variety, 176
flag-manifold, 66
function field, 142, 159
fundamental representations, 32

G-morphism, 172
G-space, 172
general linear group, 137
Gram-Schmidt process, 63
Grassmannian, 59, 71, 118
Grassmannian, isotropic, 61

Heisenberg group, 50, 83, 128, 131
highest weight vector, 115
Hilbert's fifth problem, 72
holomorphically induced representation,
  106, 116
homogeneous coordinates, 70
homogeneous space for G, 172
homogeneous space, functions on, 97
homogeneous spaces, 59
homomorphism of algebraic groups, 146

induced representations, 105
inner automorphism, 170
irreducible, 141
irreducible components, 141
irreducible representation, 84
isomorphism, 143

Index

isotropy group, 172
isotypical part, 91

Jacobi identity, 5, 75
Jordan decomposition, 151

K-finite vectors, 120
Killing form, 15, 58

Laplacian, 101
lattices, 59
left-invariant integral, 85
linear algebraic group, 146
locally closed, 144
Lorentz group, 54

Möbius transformation, 56
manifold, 69
matrix group, 92
maximal compact subgroup, 63, 120
maximal torus, 67, 180
metaplectic group, 130
metaplectic representation, 130
module for a Lie algebra, 7
MONSTER, 44
morphism, 143
multiplicative group, 147

nilpotent, 151
nilpotent Lie algebra, 8
Noetherian topological space, 141

one-parameter subgroup, 73, 182
orbit, 172
orders of finite Chevalley group, 40
orders of finite twisted groups, 42
orders of the finite Suzuki and Ree groups, 44
orthogonal group, 72, 137
orthogonality of characters, 96
oscillator representation, 129

parabolic subgroups, 117, 179
Peter-Weyl theorem, 91
Plücker embedding, 118
Plancherel theorem, 122
polar decomposition, 63
prevariety, 158
principal series, 125
projective, 160
projective space, 70, 105, 117, 159

quantum groups, 113
quasi-projective, 160
quaternions, 53
quotient of G by H, 173

radical, 154
Radon transform, 103

reduced echelon form, 64
reductive, 154
Ree groups, 43
regular, 142
representation of a Lie algebra, 7
representative function, 92
Riemann sphere, 55
ringed space, 157
root datum, 183
root system, 14, 183

Schur's lemma, 84
self-dual, 55
semisimple, 151, 153, 154
semisimple group, 58
semisimple part, 152
separability, 164
Siegel generalized upper half-plane, 62
simply connected covering group, 58, 76
singular locus, 164
smooth, 164
soluble Lie algebra, 9
solvable, 154
special linear group, 137
special orthogonal group, 137
spherical harmonics, 100, 105
spin representation, 119, 132
structure sheaf, 157
Suzuki groups, 43
symmetric space, 61
symplectic group, 137

tangent bundle, 162, 163
tangent space, 72
tangent vector, 162
tensors, 110
torus, 179
twisted groups, 41

unipotent, 151, 153, 154
unipotent part, 152
unipotent radical, 154
unitary group, 73
unitary representation, 84
universal enveloping algebra, 25
upper half-plane, 59, 125

variety, 159, 160
vector bundles, 105
Verma module, 27
Verma modules, dual, 122

weight vectors, 115
Weyl denominator formula, 88
Weyl group, 16, 66
Weyl's character formula, 30
Weyl's dimension formula, 31

Zariski topology, 140